Understanding Bridge Collapses

Understanding Bridge Collapses

Björn Åkesson

Consulting Engineer, Fagersta, Sweden

Taylor & Francis
Taylor & Francis Group
LONDON / LEIDEN / NEW YORK / PHILADELPHIA / SINGAPORE

Taylor & Francis is an imprint of the Taylor & Francis Group,
an informa business

©2008 Taylor & Francis Group, London, UK

Typeset by Charon Tec Ltd (A Macmillan Company), Chennai, India
Printed and bound in Great Britain by Anthony Rowe
(A CPI-group Company), Chippenham, Wiltshire

Published by: Taylor & Francis/Balkema
                P.O. Box 447, 2300 AK Leiden, The Netherlands
                e-mail: Pub.NL@tandf.co.uk
                www.balkema.nl, www.taylorandfrancis.co.uk,
                www.crcpress.com

British Library Cataloguing in Publication Data
A catalogue record for this book is available from the British Library

Library of Congress Cataloging in Publication Data

Understanding bridge collapses / edited by Björn Åkesson.
     p. cm.
  Includes bibliographical references and index.
  ISBN 978-0-415-43623-6 (hardback : alk. paper)
  ISBN 978-0-203-89542-9 (ebook)
  I. Bridge failures.
  I. Åkesson, B. (Björn)

TG470.U53 2008
624.2–dc22
                                            2008011017
ISBN13   978-0-415-43623-6 (Hbk)
ISBN13   978-0-203-89542-9 (eBook)

Contents

Preface and Acknowledgement   vii
Introduction   ix
List of symbols   xiii

  1. Dee Bridge (1847)   1
  2. Ashtabula Bridge (1876)   17
  3. Tay Bridge (1879)   33
  4. Quebec Bridge (1907)   53
  5. Hasselt Bridge (1938)   79
  6. Sandö Bridge (1939)   89
  7. Tacoma Narrows Bridge (1940)   97
  8. Peace River Bridge (1957)   115
  9. Second Narrows Bridge (1958)   123
10. Kings Bridge (1962)   129
11. Point Pleasant Bridge (1967)   139
12. Fourth Danube Bridge (1969)   149
13. Britannia Bridge (1970)   155
14. Cleddau Bridge (1970)   171
15. West Gate Bridge (1970)   179
16. Rhine Bridge (1971)   193
17. Zeulenroda Bridge (1973)   201
18. Reichsbrücke (1976)   213
19. Almö Bridge (1980)   223
20. Sgt. Aubrey Cosens VC Memorial Bridge (2003)   235

Literature   247
Picture and photo references   257
Index   263

Preface and Acknowledgement

It was during my research studies in early 1990's, at Chalmers University of Technology in Gothenburg, Sweden – writing a thesis on the fatigue life of riveted railway bridges – that my interest in the load-carrying capacity and life-span expectancy of bridges (preferably in steel) was awakened. My tutor (and mentor) Professor Bo Edlund, had for many years used different failure cases in his lectures to exemplify different complex phenomena, and these lectures further stimulated my interest. When I became a lecturer in 1994, it became natural to continue the work and the ideas of Professor Edlund in the lectures that I gave myself. In my research studies (where I carried out extensive fatigue tests on riveted railway stringers), and during the years after becoming a PhD (besides being a lecturer at Chalmers, also working part-time as a consulting engineer), I had the opportunity to take part in field studies and investigations of some 20 old riveted railway bridges. In these projects the focus was on the load-carrying capacity and remaining fatigue life of the individual bridges, and the knowledge from these investigations was also used in my lectures. In order to support my lectures in other topics than fatigue I had to find other relevant material about different failure cases (to support the presentation of the different phenomena), beside these already used by Professor Edlund that is. It is from this collected material that I have chosen the 20 different land-mark accidents that are presented in this book – they represent for me (and also, I believe, for the students during my lectures) "aha experiences", where not only different phenomena are made visible and clear, but also where the knowledge about complex structures such as bridges has been improved. These failure cases (some well-known and some less known) – starting with the Dee Bridge collapse back in 1847 – represent stepping stones that gradually have filled the gaps in the knowledge about the behaviour of bridge construction materials and built-up structural systems. Seen from this perspective it is then somewhat strange that bridges still continue to collapse – during 2007 there were not less than three major accidents that upset the world; The I-35W high-way truss bridge over the Mississippi River in Minneapolis, the suspension bridge over the Hau River in the Mekong Delta City of Can Tho in Vietnam, and the bridge over the Juantuo River in Fenghuang in the Chinese province of

Hunan. There is definitively still much to learn from bridge failures, as these continue to surprise and confuse (and possibly also haunt) the engineering community. Not only students, but especially contractors, practicing engineers and bridge owners continuously have to learn from the mistakes made as bridge collapses unfortunately continue to occur. As a former researcher I must say that there exists a great discrepancy between the in-depth knowledge – presented by learned scientists in journals and in papers at engineering conferences – and the actual reliability of existing bridges or bridges who are to be built. Therefore the purpose of this book has not been to describe the different failure cases in a way that suits my former fellow researchers, but instead present and discuss the issues from the perspective of a structural engineer, as the knowledge should not be hidden behind too much theoretical discussion. I have not been able to avoid theory and expressions altogether, but at least they are presented in such a way that it should be possible to understand and follow the discussion. The main purpose of this book is for the reader really to learn from the mistakes made, and be able to understand and visualize the different phenomena which, at the time of the collapse, were not fully understood. In many of the failure cases alternative solutions of how to improve the bridge designs are also presented.

Finally, I would like to address my warmest thanks to Professor Emeritus Bo Edlund, who has provided valuable comments on some of the chapters, and for being my guiding star over the years. And speaking about guiding stars, and sources of inspiration, I must also take the opportunity to acknowledge the professors John W. Fisher, USA, and Manfred A. Hirt, Switzerland, who really inspired me during my doctoral studies. Jan Sandgren and Associate Professor Mohammad Al-Emrani have both helped me by providing material as well as encouraging me in my work – I am very grateful and indebted to their help.

April 2008
Björn Åkesson

Introduction

In a bridge project the main focus for a designer lies in providing an adequate load-carrying capacity based on the assumed loads and the strength of the material to be used, and in this process lies a certain anxiety that errors in calculation perhaps could lead to failure of the entire bridge. However, errors in calculation very seldom is the main reason for a collapse – small errors is counterbalanced by safety factors both on the live load as well as on the strength of the material; in addition there is also an extra inbuilt safety in the statical system where the assumptions made normally are on the safe side (as an exception to this conclusion, the failure of the Quebec Bridge could be mentioned – see Chapter 4 – as it was partly due to an error in calculation and the fact that safety factors were more or less omitted). Instead it has been found, when bridge failures are studied, that the main cause by far is scour

(Photo: Swedish National Road Administration. With kind permission of Lennart Lindblad)

(i.e. when fast-flowing water undermines the foundation of piers and abutments) and/or the accumulation of ice or debris accumulating on to the bridge, applying horizontal pressure to the same.

However, even though scour and damming are responsible for most of the bridge failures (about 50%) no actual case in this book exactly matches this cause (yet the Peace River Bridge failure comes close – see Chapter 8). Instead I have come to concentrate on failure causes where a designer normally should be more in control (earthquake damages have for the same reason also been left out). The failure causes presented are: insufficient strength of the material used, lack of inspection and maintenance, hit damages from traffic, fatigue and brittle fracture, buckling, wind loading, aerodynamic instability, fire, inadequate anchorage capacity.

Finally, I will take the opportunity to comment upon a very interesting hypothesis which was presented by two British scientists in 1977, P.G. Sibly and A.C. Walker. They suggested that these five failure cases in history represented a paradigm shift with respect to the basic knowledge and understanding of bridge structures:

- Dee Bridge   1847
- Tay Bridge   1879
- Quebec Bridge   1907
- Tacoma Narrows Bridge   1940
- The Box-Girder Bridge Failures 1969–1971

Each of these failure cases has given its unique contribution to the general knowledge of the statics and dynamics of bridges, and with a 30-year interval, and this is the interesting observation. One engineering generation develop and improve a particular bridge concept, making the most use of its possibilities, but not really understanding the limitations with respect to a certain phenomenon or material strength. When this bridge type fails, the next generation of engineers introduces a new bridge concept, while the old generation stand back. First this new concept is developed with great caution – knowing what happened with the old one – but soon the concept is as well stretched beyond the bounds of possibility. And so it continues.

In the 1990's the American Professor Henry Petroski suggested that stay-cable bridges possibly might be the next bridge type that would fail, following the 30-year rule to come in the early 2000's. His suggestion was caused by the rather fast development with respect to maximum span length. In 1995, Pont de Normandie over the River Seine in France, the world's longest stay-cable bridge, with a main span of 856 metres, was built, surpassing the reigning stay-cable bridge in length with more than 40%. Four years later, in 1999, the

Tatara Stay-Cable Bridge in Japan was completed, having a main span of 890 metres. But fortunately, the beginning of the new millennium started without any reports of stay-cable-bridge failures, so perhaps the engineers of today have learnt from the failures of the past (or listened to the warnings). However, as bridges unfortunately still continue to collapse – for other reasons than new concepts being pushed too far – it is my strong belief that bridge engineers (active in their profession) and especially students (the bridge engineers to come, with or without presenting new structural concepts) need to learn from the mistakes made in the past.

The Author

List of symbols

$+$	tension
$-$	compression
β_A	the ratio between effective and gross area
γ_G	partial factor (load effect)
λ	slenderness parameter
$\overline{\lambda}_p$	slenderness parameter (plate buckling)
ν	poisson's ratio
ρ	reduction factor
σ	applied nominal stress
σ_{cr}	critical buckling stress
σ_{Euler}	critical buckling stress according to the Euler theory
σ_r	stress range
χ	buckling reduction factor
A	cross-sectional area
A_d	cross-sectional area of diagonals
A_{eff}	effective net area
a	plate length
a	crack length
b	plate width
b_{eff}	effective breadth
CVN	Charpy V-notch energy
c/t	slenderness ratio
D	weight
d	length of diagonals
E	modulus of elasticity (Young's modulus)
E	loss of swinging height
E	East
e	eccentricity
f	natural frequency
f	rise of arch
f	factor taking crack length and geometry into account (FM analysis)
f_y	yield strength

G	weight
g	weight per meter
H	horizontal force
H_{cr}	critical buckling force
h	depth of the web
I	second moment of area
ITT	impact transition temperature
i	radius of gyration
K	stress intensity factor
K_c	fracture toughness
k	buckling coefficient
k	column effective length factor
L	span length
L	contribution length
L	length of member
L_b	arch length
L_{cr}	buckling length
l_c	buckling length
M	bending moment
M	concentrated mass
M_{sd}	design bending moment
m	number of half-sine waves (longitudinal direction)
m_r	bending moment range per unit length
N	normal (axial) force
$N_{b.Rd}$	design axial force resistance
$N_{b.sd}$	design axial force
$N_{c.Sd}$	design axial force
$N_{c.Rd}$	design axial force resistance
P	load
P_{cr}	critical buckling load (axial load)
P_r	load range
P_{ult}	ultimate failure load
Q	weight
q	evenly distributed load
q_{cr}	critical buckling load
R	reaction force
r/t	slenderness ratio
t	plate thickness
t_f	flange thickness
W	West
W	width
$y_{n.a.}$	position of the neutral axis

Chapter 1

Dee Bridge

The collapse of the Dee Bridge in May 1847 cannot be described without first mentioning some few words about the famous Ironbridge. If the Dee Bridge represented the bitter ending of the use of cast iron in railway bridge engineering (at least for spans of 10–12 metres and longer) then the Ironbridge – with its optimal semicircular arches – represented the grandiose beginning (however, being just a road bridge, not intended for any railway traffic).

The Ironbridge, over the river Severn at Coalbrookdale in Shropshire, about 40 kilometres northwest of the City of Birmingham, was completed as early as 1779 (Fig. 1.1). The bridge, which replaced the ferry that transported people and horse-drawn carts over the river, had the inspiration coming from the classical stone arch bridges with respect to its shape. The choice of an arch for the first ever built cast-iron bridge was not only due to the inspiration from the stone bridges, but also because of the obvious requirement regarding free sailing height for the river traffic. As the

Fig. 1.1 The famous Ironbridge over the river Severn at Coalbrookdale, which was built in 1779, and is still standing intact today. (Cornell: Byggnadstekniken – metoder och idéer genom tiderna)

material was both new and untried there was no knowledge of how to join the cast-iron segments together, so the inspiration had once again to come from another direction, this time timber engineering – tenons and plugs had to be used not knowing of any other possible technique (the use of rivets in order to join iron pieces together was introduced first in the beginning of the 1800's). The splendid qualities of this new material was proven in 1795, when a spring flood damaged the major part of the stone bridges in the area while the Ironbridge was left totally undamaged (it did not dam up the water due to its open and permeable structure).

After this rather successful introduction of cast iron as a structural material for bridges, also shorter *girder* bridges were introduced. During the first part of the 1800's simply supported girders became the most common bridge type for the passage of small streams and canals, and also as shorter viaducts over roads. The cast-iron girder bridge had then also outdone the timber bridge for railway purposes (a cast-iron girder had a markedly higher load-carrying capacity and stiffness, and also less probability of being damaged or destroyed due to a fire). In comparison to the arch bridge – which was abandoned for railway purposes when the girder bridge was introduced – the girder bridge did not require any large construction depth upwards in the vertical direction (as well as downwards), which made it very suitable as a bridge type for the railways.

The span length for girder bridges increased successively, but was restricted by the maximum possible casting length ($\leq$12 m). By splicing two or more elements together, it was, however, possible to bridge over increasingly longer spans. In 1846 a cast-iron girder bridge was built that was to become the longest ever, surpassing all the other simply supported girder bridges with respect to the span length, namely the railway bridge over the river Dee just outside the town of Chester. The railway line between Chester and Holyhead (out on the island of Anglesey) was built to accommodate for the traffic by ferries to and fro Ireland (Fig. 1.2).

When the task of bridging the river Dee outside Chester was commissioned to Robert Stephenson – the famous railway engineer who a couple of years later constructed the monumental Britannia Bridge over the Menai Strait sound (between the mainland of Wales and the island of Anglesey) – he chose to limit the number of piers in the water in order to bridge the wide river in only three leaps, equally long (Fig. 1.3).

The structure became a 99.6 metres long double-track bridge, having three simply supported girder spans of 33.2 metres each. In the transverse direction the bridge was having four parallel, mono-symmetrical cast-iron girders, with the sleepers directly supported on to the lower flanges of the I-girders (Fig. 1.4).

Never before had a girder bridge having such long spans been built, and in order to carry the heavy loading from the railway traffic the cast-iron girders were reinforced with wrought iron tension bars. These bars were positioned in such a way that they followed the tension flow in a simply supported girder subjected to bending (Fig. 1.5).

The girder was spliced in two positions (due to the limitation in length at the casting), and therefore the girder was strengthened at the top in those positions. As the girder was also very narrow in depth in relation to the girder length (it should be noted that the drawing is a little out-of-scale – the girder depth should be smaller), one had also chosen to anchor the tension bars above the girder ends in order to increase their efficiency (less inclination would cause a higher force in the bars for the same loading). As a global system the girder is statically determinate, and as such relatively easy to analyse, however, the inner system is rather complex. The parts could be separated from each other, and then we have the simple girder and a Queen Post truss (Fig. 1.6).

Fig. 1.2 Great Britain and Ireland. The location of the Dee Bridge, just outside the town of Chester, is marked with an x.

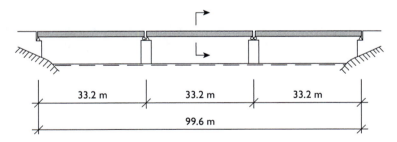

Fig. 1.3 Elevation of the Dee Bridge – three simply supported girder spans of 33.2 m each, giving a total bridge length of 99.6 m.

If the Queen Post truss had been a completely separated system, the bars would then contribute to the load-carrying capacity in such a way that a part of the load would be transferred from the chain links in the centre to the anchorage at the ends. However, as the bars are completely integrated with the girder (through connecting pin bolts), the action of the same becomes slightly different. The horizontal bar can be assumed to interact with the girder as it adds extra area to the tension flange in the centre part. The inclined bars, however, could be questioned with regard to their efficiency, at least

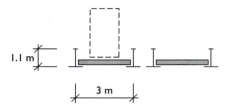

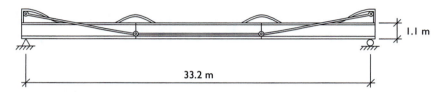

Fig. 1.4 Cross-section of the Dee Bridge – four parallel 1.1 m deep mono-symmetrical I-girders, giving a double-track bridge.

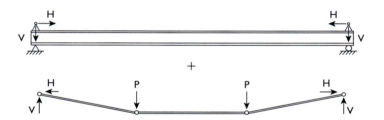

Fig. 1.5 Each girder was reinforced with wrought iron tension bars.

Fig. 1.6 The simply supported girder and the Queen Post truss system separated from each other.

with respect to the ability to transfer load to the girder ends. As the Queen Post truss is a self-anchored system (to the upper part of the girder that is), an eccentric and compressive force is introduced that counterbalances the contribution to the bending moment resistance of the girder from the inclined bars – the eccentric moment has the same sign as the bending moment produced by the vertical loading, and thus counteracts any lifting effect (see more about this action in the discussion further on regarding a possible pretensioning of the bars). As a conclusion it could be stated that the tension bars do contribute to the load-carrying capacity in such a way that the horizontal bar increases the *bending moment capacity* in the centre part, while the inclined bars are not so efficient with regard to the bending moment capacity, but at least they increase the *shear force capacity* of the girder in the end zones.

Due to the mono-symmetry the girder will under bending experience much higher compressive stresses (in the upper part) than the tensile stresses (in the lower part), however, this was also the intention when the girder was designed. From tests the relation between the tension and compression strength of cast iron had been found to be close to the relation of 3:16, i.e. that the tension strength was slightly less than one fifth of the compression strength. An optimal shape for the I-girder of the Dee

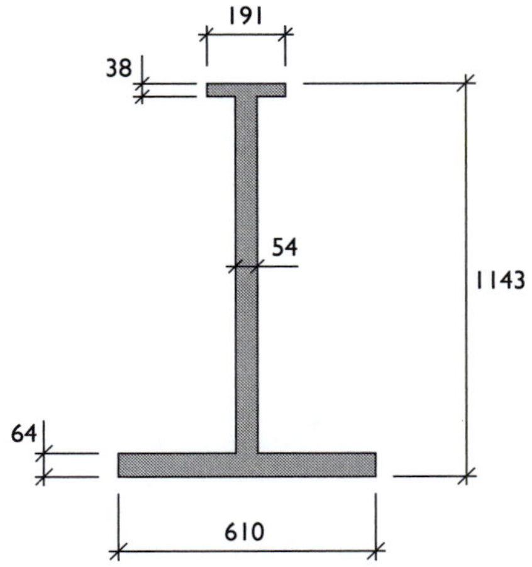

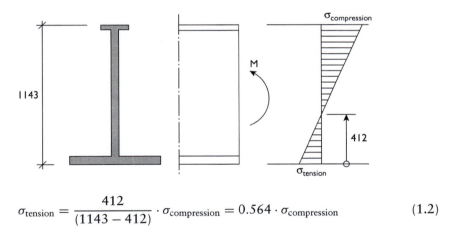

$$A_{\text{flange}}^{\text{tension}} = 610 \cdot 64 = 39040 \, \text{mm}^2$$

$$A_{\text{flange}}^{\text{compression}} = 191 \cdot 38 = 7258 \, \text{mm}^2$$

$$\Rightarrow \frac{A_{\text{flange}}^{\text{tension}}}{A_{\text{flange}}^{\text{compression}}} = \frac{39040}{7258} \cong \frac{16}{3} \qquad (1.1)$$

Fig. 1.7 Cross-section dimensions of the mono-symmetrical I-girder. The lower flange was increased in relation to the strength of cast iron in compression relative to the strength in tension, in order to make up for the difference in strength.

$$\sigma_{\text{tension}} = \frac{412}{(1143 - 412)} \cdot \sigma_{\text{compression}} = 0.564 \cdot \sigma_{\text{compression}} \qquad (1.2)$$

Fig. 1.8 The relationship between the maximum tension stress and the maximum compression stress for pure bending of the *unreinforced* girder.

Bridge – according to the designers of the time – was consequently a profile where the tension flange area was increased in relation to the compression flange with the inverse proportion (Fig. 1.7, Eq. 1.1).

However, something that was not taken into consideration was the fact that the neutral axis of this mono-symmetrical cross-section does not fall in between the flanges with the same relation (i.e. 3/16 of the girder depth). As the neutral axis of this

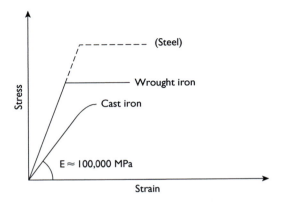

Fig. 1.9 The principal stress/strain-relationship (in tension) for cast iron in comparison to wrought iron (and steel). Cast iron is very brittle, having limited plastic deformation capacity. For one and the same stress level cast iron will deform twice as much as wrought iron.

cross-section (without tension bars) is found at 412 millimetres up from the lower face of the tension flange, the stress relation becomes altered (Fig. 1.8, Eq. 1.2).

The maximum tension stress at bending will be 56.4% of the maximum compression, instead of the intended relation 3/16 (18.75%). Thus the maximum tension stress at bending was very much underestimated, despite having taken the difference in strength into consideration by shaping the I-profile with different flange sizes. However, as the girder – in addition to the mono-symmetrical shaping – also was strengthened by the applied wrought iron tension bars, there is perhaps reason to believe that the load-carrying capacity was heavily enlarged, especially as the wrought iron tension bars have twice as high modulus of elasticity as the cast-iron girder (wrought iron has a modulus of elasticity of about 200,000 MPa, i.e. in the neighbourhood of steel, while cast iron has a value of something like 100,000 MPa) (Fig. 1.9).

The girder was in the mid-span region part reinforced in the lower part with four wrought iron tension bars – each being $32 \times 152 \, \text{mm}^2$ in size – two on each side of the girder web (with a small free spacing in order to pass the upper flange for the anchorage at the ends). The stress relation is changed in a positive direction due to the fact that the neutral axis is lowered, however, not as much as one perhaps would expect, so the tension stresses are still being underestimated (Fig. 1.10, Eq. 1.3).

With respect to the load-carrying capacity – based on the section modulus for the lower part of the girder subjected to tension – it will increase with approximately 25% for the tension bar reinforced cross-section in comparison to the simple girder. However, as the magnitude of the tension stresses still is being underestimated the load-carrying capacity will still be insufficient. An assumed load test using the design load of that time (a 30 ton heavy and nine metre long locomotive) – three of those positioned close to each other on one of the girder spans – would result in a stress level on the tension side of about 80 MPa, which is on the level of the unreduced strength (read: the strength without any safety factors) for cast iron of *very* high quality. A load test was also performed, however, using less heavy locomotives, and that was lucky,

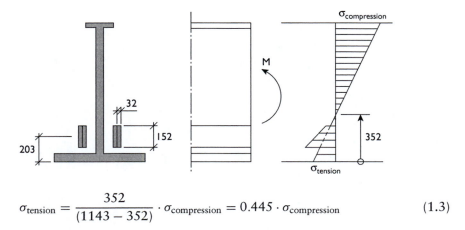

$$\sigma_{\text{tension}} = \frac{352}{(1143 - 352)} \cdot \sigma_{\text{compression}} = 0.445 \cdot \sigma_{\text{compression}} \qquad (1.3)$$

Fig. 1.10 The relationship between the maximum tension stress and the maximum compression stress for pure bending of the *reinforced* girder.

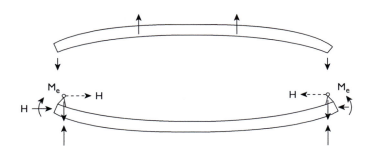

Fig. 1.11 The up-lift and the eccentric attachment at the ends balance each other out.

otherwise one would have registered – besides the large deflection (which perhaps still was observed) also a collapse already at the time of the load test.

There is also perhaps reason to believe that a possible pretensioning of the tension bars (e.g. by attaching the bars in a heated condition, and thus introducing a contractive force after cooling), would reduce the tension stresses arising due to the in-service train loading. However, as has been discussed before (with respect to the Queen Post truss and its influence as an integrated system), the "up-lifting" effect will be counterbalanced because of the eccentric anchorage at the ends (Fig. 1.11).

And even if the pretensioning would be more efficient for the response at in-service loading – e.g. by having the horizontal bar running all the way along the girder and attached at the lower region at the ends – the maximum load-carrying capacity in the ultimate limit state would not be affected as the forces produced by the pretensioning (tension in the bar and compression in the girder) are inner self-balancing forces.

With or without any efficient pretensioning of the bars, the bridge was taken into service in November 1846, but was already after some weeks taken out of service,

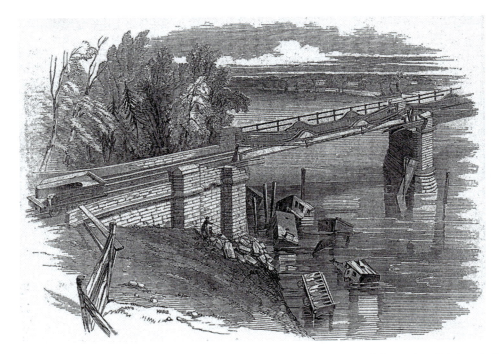

Fig. 1.12 As the train passed the last span over the river Dee, one of the girders failed, and all of the carriages followed the girder into the river below. (http://en.wikipedia.org/wiki/Dee_bridge_disaster)

as a crack had been found in the lower flange of one of the girders. At the repair it was found that the tension bars had not been properly installed, thus making the girder becoming overloaded when trains were passing. However, it was at the same time an apparent proof of the reinforcing effect coming from the tension bars, which saved the bridge from a complete collapse by coming into action after the lower flange of the girder had broke. After repair of this damaged girder, and inspection of the remaining others, the bridge could once again be taken into service. The bridge was thereafter performing without any problems; however, only some six months after the bridge had been taken into service, the bridge collapsed during the passage of a train. The train had left the station in Chester at 6.15 P.M., 24 May 1847 – on its way to Ruabon, approximately 25 kilometres southwest of Chester – and arrived at the Dee Bridge site after just some few minutes. The locomotive and the carriages did pass the two first girder spans without any indication of anything being wrong. However, as the train set was passing the last span, the driver felt the carriages sink beneath him. He pulled full throttle for maximum steam pressure and the locomotive and the tender (the carriage behind the locomotive which carries fuel and water) succeeded to reach the shore behind the abutment, but all the remaining carriages fell into the river. It was the outermost girder – of the two in this last span – that had failed (Fig. 1.12).

Five people died in the collapse, and the remaining passengers (about 30) were all more or less seriously injured. The driver did show a great presence of mind as he continued to the nearest train station in order to alarm the next coming trains.

Eyewitnesses distinctly talked about a crack that opened up from the lower part of the girder as the bridge span was loaded by the train, and that the last carriage in the train set fell into the river first, followed by the other carriages.

The same morning a 12.5 centimetre thick layer of ballast had been added to the bridge (supported by a wooden deck on top of the sleepers) in purpose of protecting the bridge against any fire hazard (a recent bridge fire – due to hot ashes/sparks – had attracted the attention to this potential risk). This additional loading of about 18 tons on each bridge span was apparently judged as something that the girders could carry, however, it also meant that the mean stress level was increased by some 8–9 MPa, which is a markedly stress raise to an already heavily strained girder. As the fatal train to Ruabon was to pass the bridge later the same day, the total stress loading became too much for the girder to carry, and the tension strength of the same was exceeded.

Over the years a number of different explanations have been put forward to why the bridge did collapse. First and foremost it was focused – and quite rightly so – on cast iron as a suitable (or unsuitable) material for structures subjected to tension. Its brittle characteristics in combination with the low tensile strength, and the presence of inner defects and weaknesses, made structural engineers at large – many of those already before the collapse – recommend the more strong and ductile wrought iron material. Cost and supply were something that already in the 1800's governed the choice of a structural material, and that made the cheaper and easier to produce cast iron more competitive than wrought iron. Cast iron had also the advantage of a superior corrosion resistance.

Stephenson himself stood up and defended the load-carrying capacity of the bridge, and he claimed *derailing* as the direct cause of the collapse – the train had, according to him, hit the girder and thus initiated the fracture. This explanation can not be ruled out; however, the probability of derailing and the character of the fracturing (as described by the eyewitnesses) speak against such a cause. In addition, an extra set of rails was present in the bridge protecting against derailment.

Another explanation, which in modern time has been suggested, is *instability* – namely that the thin and narrow upper flange should have initiated lateral/torsional buckling due to the high normal compressive stresses in the upper part of the girder. There are, however, two things that speak against such an instability phenomenon occurring – even though the girder at first sight seem entirely to lack any lateral stability with respect to the upper flange dimension. First and foremost, the point of time for the collapse – six months after the bridge was taken into service – is a proof in itself that instability cannot be the cause of the failure. If a girder is unstable (against L/T-buckling or normal stress buckling) then it will also buckle already at the first loading. Second, the girder – supported as it is on the wide flange – is comparable to a T-profile put upside down, which is stable in itself. Would the upper flange – subjected to compression as it is – tend to buckle in the sideways direction, then the lower flange would counteract such a deformation. One could say that the web plate is fixed at its base to the lower flange, and that the web plate is working as a restraining spring to the upper flange. In addition, the web plate is in itself sufficiently compact (read: with its thickness of 54 millimetres not being enough slender), which makes it highly unlikely

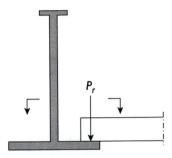

Fig. 1.13 The sleepers introduce an eccentric loading to the lower flange plate.

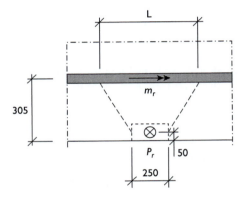

Fig. 1.14 The approximate position on the lower flange of the resulting load from the sleepers.

for any normal stress buckling to occur. And with respect to a possible L/T-buckling risk there is also a stabilizing effect coming from the fact that the load is applied on the lower flange, and not on the upper.

One of the most recent ideas presented is failure because of *fatigue*, due to the eccentrically applied loading on the lower flange from the sleepers (Fig. 1.13).

In the figure above the wooden decking on top of the sleepers is missing – where the ballast was put – and also the tension bars close to the web (the position of the same explains why the sleepers could not be put directly against the web plate).

For a varying load P (here with an index r for range), positioned approximately 50 millimetres in from the flange edge, it is possible to determine the local bending moment in the lower part of the girder (Fig. 1.14).

If we consider that the maximum axle loading of that time (being 8–10 tons), we are able to calculate the cyclic load (read: the load range). We must also take into consideration the load distribution effect in the longitudinal direction coming from the rail (Fig. 1.15, Eq. 1.4).

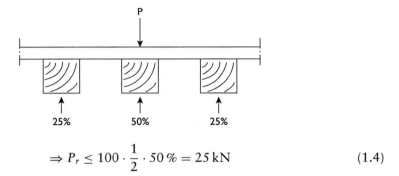

$$\Rightarrow P_r \leq 100 \cdot \frac{1}{2} \cdot 50\% = 25\,\text{kN} \tag{1.4}$$

Fig. 1.15 The load distribution among adjacent sleepers.

We assume a dispersion of 30° in order to determine the contribution length (Eq. 1.5):

$$L = 250 + 2 \cdot (305 - 50) \cdot \tan 30 = 544\,\text{mm} \tag{1.5}$$

As we now finally calculate the actual bending moment, we neglect any possible (but negligible) overlapping effects at the edges of this contribution length, coming from adjacent sleepers being also loaded (Eq. 1.6):

$$m_r = \frac{25 \cdot (305 - 50)}{544} = 11.7\,\text{kNm/m} \tag{1.6}$$

In conclusion we obtain the maximum stress range (in the weakest part, which is in the web plate) (Eq. 1.7):

$$\sigma_r = \frac{11.7 \cdot 10^{-3}}{\left(\dfrac{1.0 \cdot 0.054^2}{6}\right)} = 24.1\,\text{MPa} \tag{1.7}$$

We have – in this calculation of the maximum stress range – not added the stress coming from the load that has to be suspended to the web; however, a quick check shows that this additional stress is more or less negligible (Eq. 1.8):

$$\sigma_r^{\text{tension}} = \frac{25 \cdot 10^{-3}}{0.054 \cdot 0.544} = 0.85\,\text{MPa} \tag{1.8}$$

The suggested fracture scenario was that an initiated fatigue crack, in the transition area between the flange and the web in the lower part of the girder, gradually would have propagated to a certain critical length (with respect to brittle fracture) and thus caused the sudden collapse. However, we can see from the calculations above that the *maximum* stress range not the least is of such magnitude that the fatigue limit for cast iron would be exceeded, and this is still the fact despite that cast iron is having blisters and other impurities, which makes it weaker (and more brittle) than modern steel of today. There is also an additional stress raising factor to consider in the affected area, and that is the especially designed fillet transition between the flange and the web (Fig. 1.16).

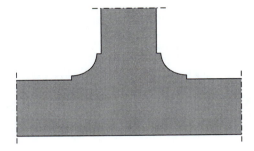

Fig. 1.16 The fillet transition between the flange and the web acts as a stress raiser as this area is not smooth.

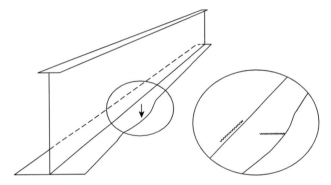

Fig. 1.17 Possible fatigue cracking locations due to local bending of the flange plate.

This decorative shaping is raising the local stress level, but not as much as for a similar welded detail (in steel), which have a fatigue limit that exceeds the stress range level calculated here. We must also not forget that the calculated stress range is for an assumed *maximum* axle weight, which is rather unlikely to occur, at least with regard to any large number of cycles. During the six months the bridge was in service the local stress reversals produced were not of any significant magnitude, and those stress cycles that did occur were still having a fairly low number of repetitions (due to the limited period of time together with the relatively low number of train passages per day). Consequently we could rule out the possibility of any fatigue cracking to occur. And if we – despite the fact that no fatigue cracking could have taken place – still imagine us a crack in the upper part of the fillet being initiated, then this crack would be *horizontal*, i.e. parallel to the normal stress direction due to the global bending of the girder. Such a surface crack – horizontal and single sided (i.e. not going through the thickness of the web plate) – does not constitute a direct threat to a possible brittle fracture risk, which transverse cracks do. For local bending of the flange there is, in a way, also a possible risk for a transverse crack to be initiated, however, the bending stresses here are much lower than those that arise in the web-to-flange transition, so for the case of the Dee Bridge a transverse crack is even less probable than a horizontal (Fig. 1.17).

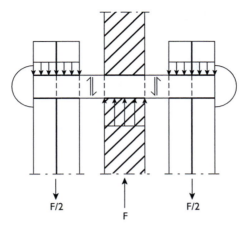

Fig. 1.18 Through shear in the pin bolts the elongation of the lower part of the girder (when subjected to bending) is restrained by the wrought iron tension bars.

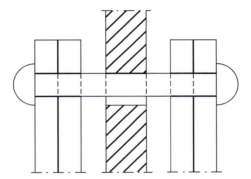

Fig. 1.19 Due to the repeated loading the bolt hole in the girder web becomes over time elongated (ovalized).

What could instead be a possible cause of failure is a successive *ovalization* of the hole for the pin bolts that join the tension bars together (on either side of the web plate) to the girder. As the girder becomes subjected to bending it will elongate in the lower part, and through this elongation the horizontal tension bars will be forced to interact with the girder (visualized in the figure below for an imagined force F) – the tension bars are consequently acting as a constraint against this elongation (read: they unload the girder by supplying extra tension flange area) (Fig. 1.18).

Because of the relative hardness of the tension bars and the pin bolt (in comparison to the softer cast-iron material in the girder, see Fig. 1.9) it will in time develop a localized deformation in the contact pressure area between the pin bolt and the web plate of the girder (the contact pressure here is also the largest). This will lead to an ovalization of the hole (read: remaining plastic deformation) – the hole has quite simply been elongated through wear and excessive local contact pressure (Fig. 1.19).

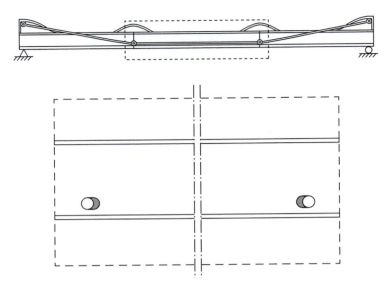

Fig. 1.20 Elongated holes in the cast-iron girder web plate.

An elevation of the girder (unloaded condition) having the hidden elongated holes in the web plate is shown above (Fig. 1.20).

As the holes are more or less hidden behind the tension bars (as well as behind possible washers and girder reinforcements) they could be difficult to check even at a close inspection. However, what had been discovered was that the deflections of the girders were remarkably large – at most some 14 centimetres in midspan. If the deflection is calculated for a normal train load of that time – assuming full interaction between the tension bars and the girder – then the deflection will be approximately half of this registered value, and this indicates quite strongly that the composite action more or less had been lost.

Ovalization due to repeated local pressure is something that occurs relatively often at fatigue loading tests on both riveted and bolted connections in modern tests as well, and becomes evident as the deformations (both local and global) gradually increase. Observe that it is then a question of steel against steel, not wrought iron against soft cast iron. As the tension bars of the girders in the Dee Bridge gradually became inactive – because of the ovalization – and that the already from the beginning overstressed girders (by the initial overestimation of their load-carrying capacity) in this way became more and more strained – not to forget about the additional last-minute loading from the ballast – the girder in the last span came to fracture in the lower flange in a brittle manner. An increased flexibility (due to the ovalization) will also lead to increased dynamic amplification of the loading, and this may also have contributed to the collapse. In contrast to the early fracture that took place already after some few weeks, the final fracture was too much for the tension bars to prevent alone (being made "redundant" because of the ovalization as they were). As the tension bars were not able to withstand the loading, the crack separated the girder in two halves, and then

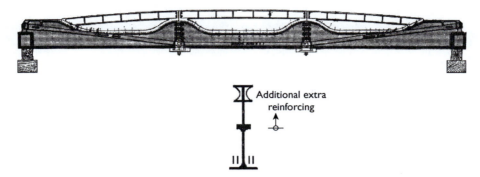

Fig. 1.21 As a consequence of the Dee Bridge collapse many existing cast-iron bridges had to be strengthened. (Berridge: The girder bridge after Brunel and others)

the Queen Post truss also lost its function. However, if the crack also had stopped in the lower flange this time (as it did in November 1846, when the first cracking was detected), then the Queen Post truss system would have been able to carry all the load, so much load-carrying capacity was the tension bars having.

The collapse of the Dee Bridge had some large repercussions. An investigation did result in both that the choice of material, and the tension bar reinforced girder system as such, was blamed for the collapse and more or less doomed this bridge type for further use. Most of the already existing cast-iron girder bridges were either replaced or strengthened. An example of the latter was the 18 metre long and also tension bar stiffened girder bridge over Trent Valley, which was strengthened by the adding of extra plates and profiles in order to increase the girder depth (Fig. 1.21).

What had been such a glorious start for cast iron as a structural material for bridges – when Ironbridge was built – ended in such a catastrophic failure when the Dee Bridge collapsed. All the limits there were had boldly been stretched – from an arch having dominating compression to a girder subjected to excessive bending, and from a simple road bridge to a heavily loaded railway bridge. A choice of the stronger and, above all, more ductile wrought iron material would definitely have saved the Dee Bridge. Steel was first introduced as a structural material in the 1860's, so that was not yet an option.

By still continuing with traffic on the bridge, after the initial incident with the girder flange fracture, really shows that there was an irrepressible pioneering spirit at that time – anything was possible, nothing was impossible. The designer had an absolute power to test any idea of his own, deciding about what material to use and what kind of structure to build.

Chapter 2

Ashtabula Bridge

On the evening of 28 December 1876, a steam engine driven passenger train of eleven cars left the city of New York, on the east coast of USA, in order to transport some hundred passengers back west for the upcoming New Year Holiday. This particular railroad was owned by The Lake Shore and Michigan Southern Railway Company, a railway company that was formed in 1869 in order to connect New York with Chicago, passing through the cities of Buffalo and Cleveland on the shore of Lake Erie (Fig. 2.1). Two o'clock in the afternoon of the next day, Friday 29 December, the train left Buffalo for its continued travel of about 280 kilometres to Cleveland, Ohio, approximately one hour behind time schedule.

Because of a heavy snow storm, an extra steam engine had to assist in pulling the train just after Buffalo. Due to the snow fall, the speed was reduced to 25 km/h as the train, around half past seven in the evening, approached the small town of Ashtabula,

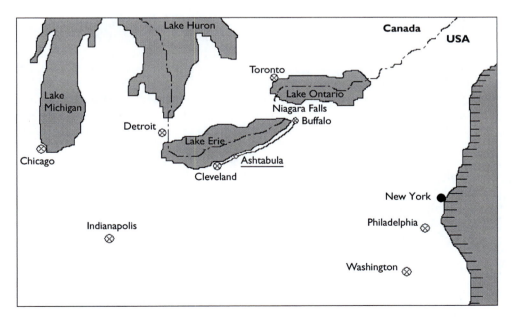

Fig. 2.1 The north-east corner of the USA and the Great Lakes region (the Lake Superior not shown).

Fig. 2.2 The Ashtabula Bridge some years before the collapse. (www.prairieghosts.com/rr_
disaster.html)

in the north-eastern corner of the State of Ohio. The visibility was more or less non-existent as the train slowed down in order to pass over the Ashtabula River Bridge, just before reaching the station. As the leading locomotive was almost at the end of the bridge, the driver heard a cracking coming from the bridge and he suddenly felt that he was driving "up-hill". As he immediately understood that the bridge was to give way, he pulled the throttle for maximum steam pressure – just as the driver did in the Dee Bridge collapse (see Chapter 1). The acceleration force forward, and the pulling force backwards from the second engine and the cars on the bridge span, made the coupling-device between the two locomotives break. The front locomotive managed to reach the west abutment on the river-bank, and the driver could in his despair and horror see how the cars on the bridge, as well as the cars still behind the abutment on the east river-bank were pulled down in to the river below. The collapse was in itself a slow and prolonged process, as the train functioned more or less like a linked chain load, which gave a certain resistance to the bridge failure. The photo above of a working train on the Ashtabula Bridge (taken some years before the collapse) does show the approximate position of the front engine as the cars behind went down with the bridge (Fig. 2.2).

A passenger on the train, Mr Burchell from Chicago, described the scenario as follows:

> "The first thing I heard was a cracking in the front part of the car, and then the same cracking in the rear. Then came another cracking in the front louder than the first, and then came a sickening oscillation and a sudden sinking, and I was thrown stunned from my seat."

One of the other surviving passengers, from one of the rear sleeping cars, Ms Marian Shepherd, gives a vivid and frightening description of the course of events:

> "The passengers were grouped about the car in twos, fours, and even larger parties. Some were lunching, some were chatting, and quite a number were playing cards. The bell-rope snapped in two, one piece flying against one of the lamp glasses, smashing it, and knocking the burning candle to the floor. Then the cars ahead of us went bump, bump, bump, as if the wheels were jumping over the ties. Until the bumping sensation was felt, everyone thought the glass globe had been broken by an explosion. Several jumped up, and some seized the tops of the seats to steady themselves. Suddenly there was an awful crash. I can't describe the noise. There were all sorts of sounds. I could hear, above all, a sharp, ringing sound, as if all the glass in the train was being shattered in pieces. Someone cried out, 'We're going down!'. At that moment all the lights in the car went out. It was utter darkness. I stood up in the centre of the aisle. I knew that something awful was happening, and having some experience in railroad accidents, I braced myself as best as I knew how. I felt the car floor sinking under my feet. The sensation of falling was very apparent. I thought of great many things, and I made up my mind I was going to be killed. For the first few seconds we seemed to be dropping in silence. I could hear the other passengers breathing. Then suddenly the car was filled with flying splinters and dust, and we seemed breathing some heavy substance. For a moment, I was almost suffocated. We went down, down. Oh, it was awful! It seemed to me we had been falling two minutes. The berths were slipping from their fastenings and falling upon the passengers. We heard an awful crash. It was as dark as the sound died away and there were heavy groans all around us. It was as dark as the grave."

Although many people were estimated to have died instantly in the crash, several passengers also perished in the fire that followed, bringing a total of 92 casualties. The severe fire was a direct result of the fact that each carriage was heated by stoves, besides being lit by oil lamps (Fig. 2.3).

The following day the charred remains of the bridge and the train lay twisted beyond recognition in the river bed below (Fig. 2.4).

A news reporter cabled the following dispatch from Ashtabula to Chicago Tribune the day after the collapse:

> "When morning came, all that remained of the Pacific Express was a windrow of car wheels, axles, brake-irons, truck-frames and twisted rails lying in a black pool at the bottom of the gorge. The wood had burned completely away, and the ruins were covered with white ashes."

Following a major tragedy of such magnitude as was the case here, a jury was immediately assembled to investigate the causes behind this bridge collapse. But before coming to their conclusions, we will temporarily stop the discussion regarding the failure and instead go back in time to the year when the bridge was built.

Originally, there was a wooden truss bridge spanning the deep ravine over the Ashtabula River – a shallow river that flows into Lake Erie, which had been given its name by the Iroquois Indians ("Hash-tah-buh-lah", meaning "river of many fish"). In order to upgrade the railway bridge by increasing its load-carrying capacity it was decided to replace this old bridge in 1865. The chief designer, Amasa Stone, decided upon a Howe truss, a very popular choice for railway bridges at that time (Fig. 2.5).

Fig. 2.3 A vivid drawing of the collapse scene (Harper's Weekly, 20 January 1877). (www.catskillarchive.com/rrextra/wkasht.html)

Fig. 2.4 The day after the collapse. The extra brick pier that is seen close to the left-hand abutment is from the earlier wooden bridge at this site. (http://home.alltel.net/arhf/bridge.htm)

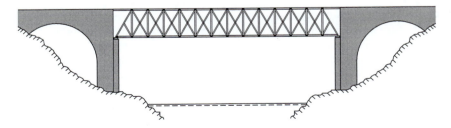

Fig. 2.5 The Howe truss bridge over the Ashtabula River, built in 1865.

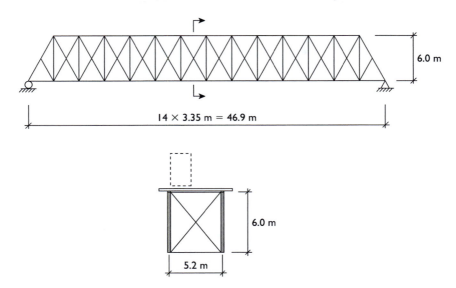

Fig. 2.6 The 46.9 m long and simply supported Howe truss bridge.

The new bridge was a 6.0 m deep all-iron truss bridge, having 14 panels of 3.35 m each, giving a total length of 46.9 m. The two parallel trusses – separated 5.2 m apart – gave place for a double track bridge (Fig. 2.6).

The original Howe truss concept – that was patented in 1840 by William Howe – used wooden members for the diagonals and the horizontal chords, and wrought iron bars for the verticals. This patented truss became very popular, as it combined many novel features. The individual parts could to a large extent be prefabricated, a need for longer spans was easily met by just adding a few more elements, and the truss could be assembled in advance and transported by train to the bridge site (however, the Ashtabula Bridge was built using a wooden falsework as a temporary support during assembly at the bridge site). The Howe truss had also the advantage that it was easy to prestress, which enabled for the (wooden) diagonals in tension to be active during loading (through a reduced compression), and it also meant a simpler connection at the joints (no need for a full positive attachment, i.e. the compression reduced the need to have a full and strong connection for tension forces). The Ashtabula Bridge followed the Howe truss concept with respect to the statical system (and also for the prestressing technique, which soon will be discussed upon), but was – as has been mentioned – an all-iron bridge, not using wooden elements for any member. The bridge was very strong and robust (at least it was considered so by Amasa Stone, the designer) and had a depth-to-span ratio less than 1:8, which could be compared to the Dee Bridge girders, having 1:30 (not taking the tension bars into consideration). The high stiffness of the truss was shown at the load-test (proof-loading) when the bridge only deflected some 15 mm under the combined loading of three locomotives.

But before having tested the load-carrying capacity at proof-loading, and before removing the falsework, the diagonals were put under compression by tightening the

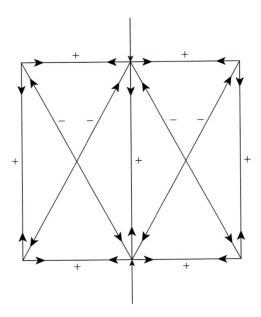

Fig. 2.7 By tightening the verticals (and by doing so, compacting the truss) the diagonals became
subjected to compression.

vertical bars, one by one. By turning the nuts at the threaded ends of the bar, tension
was induced because of the elongation, hence subjecting the diagonals to compression
due to the compacting of the truss (Fig. 2.7).

Had the truss been inner statically *determinate* (i.e. having only single diagonals in
each panel) the tightening of the verticals would then only have meant the deforming
of the horizontal chords, i.e. not having any prestressing effect at all (Fig. 2.8).

The advantage of an inner statically indeterminate truss (i.e. by having crossing diagonals) is not only the ability to prestress the diagonals, but also that there is a certain
robustness in the system allowing for alternative load-paths under external loading.
In the following simply supported two-panel truss, having crossing diagonals in each
panel, there are two alternative load-paths for the load to be transferred through the
panels during loading – one pair of diagonals in compression and one pair in tension
(Fig. 2.9).

This particular behaviour means that the truss has a certain structural integrity
(redundancy) which makes it able to withstand excessive loading – e.g., even if one of
the diagonals in a panel is lost (due to a hit damage or similar) the load-carrying ability
of the truss would still be intact, as one possible load-path would still remain. With
respect to the reaction forces (at the bearings) there is no difference between a simply
supported inner statically indeterminate truss and a simply supported inner statically
determinate truss – both structures are statically *determinate* with respect to the system
as a whole. The difference lies in the member forces, as the size of the members in an
inner statically indeterminate truss governs the distribution of the forces, whereas in an

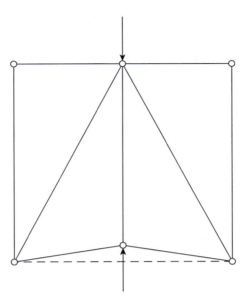

Fig. 2.8 It is impossible to induce prestressing in an inner statically determinate truss by applying a force to the verticals.

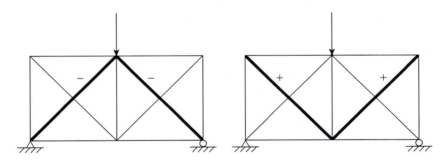

Fig. 2.9 Two separate load-carrying systems in an inner statically indeterminate truss.

inner statically determinate truss they do not. If there is an increase in a member size – in an inner statically indeterminate truss – then there is also a change in the relative distribution of forces between the individual members (the member with the increased area will be stiffer, and thus have a larger force). The same change of a member size in an inner statically determinate system will not result in any change of the inner distribution of forces between the members – the member forces are fixed by using simple equilibrium equations, and do not depend upon the member sizes. However, for a statically indeterminate system the distribution of the member forces has to be found using deformation/displacement methods. For both (simply supported) systems there is no change in the bearing reaction forces though, given that you, for example, alter one or two member sizes within the systems – both systems are, with respect to the global behaviour, statically determinate.

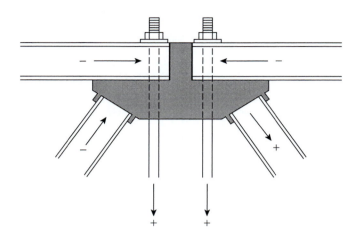

Fig. 2.10 The principal truss joint design of the Ashtabula Bridge – a cast-iron angle block in between the individual members (diagonals, verticals and horizontal chords).

As the diagonals in the Ashtabula Bridge were somewhat loosely fitted to the joints in the panels during assembly, they relied upon the prestressing (read: compacting) to function properly during loading of the truss. The tension diagonals could only carry load below the prestressing level (as a reduction of the compression), and were not able to withstand the full tension capacity of its members (due to the inadequate fastening to the joints).

There was no proper control of the prestressing level during tightening of the verticals as buckling of the diagonals did occur in some instances (the prestressing force was of course diminished in these cases). Buckling also occurred when the falsework finally was removed (then under the combined loading of dead load and prestressing force in the diagonals subjected to compression).

The joint between the diagonals and the upper and lower chords had the following principal design – a cast-iron angle block supporting, and transferring the forces between the members in the truss (Fig. 2.10).

Even though the bridge was considered to be very strong and robust, it was judged "too experimental" by another involved engineer, Mr Charles Collins, and experimental it certainly was. Both the chords and the diagonals consisted of *separate* wrought iron I-girder elements (Fig. 2.11).

Amasa Stone chose to favour his brother Andros Stone, who was the joint owner of a rolling mill (one of the earliest in USA), and therefore decided upon an all-I-girder built-up iron Howe truss (except for the verticals). The top chord consisted of five parallel I-girder elements of each 6.7 m (2 × 3.35 m) – three spliced in one joint location (at a cast-iron angle block, see Fig. 2.10), and two spliced at the next. These five girders were only held together in each panel by two through bolts. All I-girder elements had one and the same cross-section; 102 mm wide and 152 mm deep unit elements (Fig. 2.12).

Instead of using solid elements for the truss members, and adjusting the cross-sectional area depending on the maximum force to bee expected, Amasa Stone chose to work with the *number* of I-girders as the working parameter in the design process.

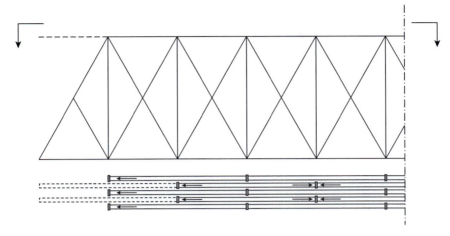

Fig. 2.11 The top (compression) chord of the Ashtabula Bridge.

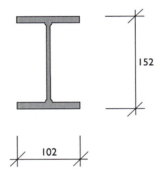

Fig. 2.12 The cross-section of the small I-girders used for the truss elements in the Ashtabula Bridge.

In every other joint in the upper chord, the I-girders were fitted to the lugs on the cast-iron angle blocks by using shims – i.e. thin plates – to fill the small gaps there during the assembly (see Figs. 2.10 and 2.11). In a sense the entire bridge resembled more a gigantic model construction kit – such as a Meccano without nuts and bolts – rather than a normal steel truss of modern standards, which is having the members properly and securely fixed to the joints. Perhaps Amasa Stone had too much of his inspiration coming from the wooden Howe trusses.

Also the transverse floor-beams (carrying the track) were of the same wrought iron I-girder profiles that were used for the truss members, and these floor-beams were just simply positioned every 1.12 m along the upper chord of the truss (Fig. 2.13).

Experimental or not, the bridge successfully came to carry the in-service train loading for more than eleven years without any major incident occurring (at least not any reported). Obviously the static strength of the bridge was adequate (even for the occasional passage of two trains passing the bridge at the same time in opposite directions).

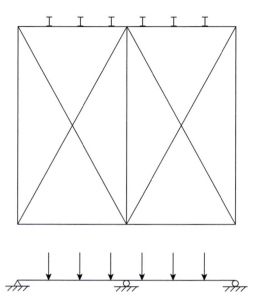

Fig. 2.13 The upper chord of the trusses had to, besides carrying the normal compressive forces due to the truss action, also – as a continuous beam – carry, in bending, the floor-beams positioned in between the truss joints.

Fig. 2.14 The cast-iron lug in the upper compression chord that was found to have been broken off (see also Fig. 2.11).

However, during the passage of the west bound train to Cleveland on the evening of 29 December 1876, the Ashtabula Bridge suddenly collapsed. An inquest was immediately summoned in order to find the answer to the collapse, and this investigation would last for more than two months. It was found that the bridge had been badly maintained, poorly inspected, and not properly designed with respect to the truss members (e.g. by applying load from the floor-beams to the upper chord between the truss joints, and by having relied upon separate elements to act as a unit), but above all, that the entire load-carrying capacity rested upon the small cast-iron lugs at the truss joint angle blocks. They discovered, after having salvaged some key parts from the wreckage of the bridge superstructure that one of the two lugs at the second upper chord joint had been broken off (Figs. 2.14 and 2.15).

There were five lugs at both ends of the upper compression chord that were having one-sided loading – the two in the second joint were having I-girder elements on either side, however, the two I-girders continuing to the abutment (see Figs. 2.5 and 2.11) are only "zero-bars" with respect to the truss action. The remaining lugs – all having

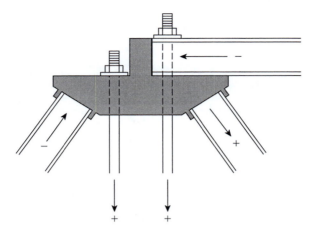

Fig. 2.15 The one-sided horizontal loading from the upper chord I-girder member to the cast-iron lugs at the ends (see also Fig. 2.10).

active members on both sides (see Fig. 2.10) – had only to transfer (by shear) the small difference in member forces between the horizontal I-girders. It was considered without any risk to choose cast iron for the angle blocks (containing the lugs), as they were to be subjected to compression only, however, the five lugs at the ends must have been somewhat overlooked in the design process. After eleven years of in-service loading one of the lugs broke, which resulted in an instant over-loading of the remaining I-girders in the chord. As the upper chord consequently became eccentrically loaded, it began to buckle outwards in the transverse lateral direction (first perhaps as separate units, and then finally as a complete "package"). There was limited capacity of the upper chord to withstand this excessive loading as the compression members of the bridge had been designed with respect to a stress criterion only, and not with respect to the risk of buckling. (Further, there is the simultaneous transverse vertical loading from the floor-beams.) One could also assume uneven load distribution in between the I-girder elements – prior to the breaking of the lug – due to difference in contact pressure to the lugs. The engine drivers had also for many years experienced a "snapping sound" when trains were passing over the bridge, indicating perhaps that even small gaps (due to the loss of shims) were present between certain compression elements in the bridge (upper chord as well as diagonals). There had also been inspection reports of misalignments – the girder elements were in many cases not meeting the lugs completely straight. Anyway, the upper chord failed in buckling, which resulted in the progressive failure of the entire second truss panel, transforming the inner statically indeterminate truss from a load-carrying structure to a complete mechanism (Fig. 2.16).

If the cast-iron lugs at the angle blocks were the weak points initiating the failure how can we explain that the bridge successfully could carry service load for more than eleven years? The static strength of the lugs could very well be adequate (given that shear was the predominant load); however, there is also fatigue to consider, even in brittle materials such as cast iron, especially for the case where the horizontal load is transferred by bending (or rather, a strut-and-tie mode of action). The shims could

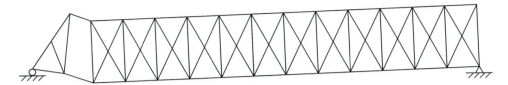

Fig. 2.16 The mechanism mode (just prior to the final collapse when the lower tension chord lost its connection to the third joint). This failure sequence explains why the driver of the first engine heard a cracking sound from the bridge behind him, and why he had the sudden feeling of driving "up-hill".

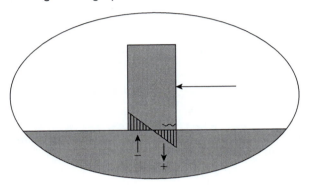

Fig. 2.17 A fatigue crack is formed perpendicular to the tensile stresses at the base of the lug.

have given uneven contact pressure zones, which could explain why some lugs were subjected to bending (and shear) rather than pure shear. When such a lug is repeatedly loaded – especially those at the ends (being loaded from only one side as they were) – a fatigue crack is eventually initiated, and this crack is then propagating horizontally (for every new loading cycle) until it reaches a certain critical length (with respect to brittle fracture of the net section) (Fig. 2.17).

The very low ductility of cast iron leads to small critical crack lengths, but still a crack has to be initiated and then propagate to give a final fracture (in a brittle manner). In addition, there was also the probability of sudden thrusts from the upper chord I-girders to the lugs if they were not exactly in close contact, and this would reduce the fatigue life of the lugs (read: reduce the critical crack length). Furthermore, at the evening of the collapse the temperature was low, which increased the brittleness (read: reduced the fracture toughness) of the cast iron material. There is one other parameter though, which is reducing the load on the lug, and that is the friction (between the I-girders and the horizontal face of the angle blocks), which is due to the pretensioning of the truss and the live load upon the floor-beams.

The inquest correctly pinpointed the broken lug as the initiating point for the collapse, but it was not until later that it was found – when examining the lug more in detail – that it contained a large flaw. An air hole (void) was found in the broken lug (Fig. 2.18).

Besides reducing the effective cross-section of the lug (not only the gross section alone, but also the net section after the initiation of the fatigue crack), the void also

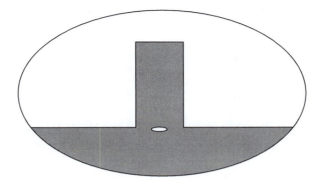

Fig. 2.18 When the broken lug was examined a large embedded air hole was found.

acts as a stress raiser due to the stress concentration effect. The transition zone between the lug and the main body of the angle block is also a region where cooling is slow after casting – in such zones there is not only the risk of voids forming (due to "suctions" because of uneven shrinkage of the metal during solidification), but also the formation of large grains, the accumulation of slag inclusions and other impurities, which all result in extra brittle behaviour in such zones. Given this knowledge, it was just pure chance that the lug did not break at first loading.

While the inquest was investigating the reason behind the collapse and who should take the responsibility, some relevant "layman questions" were raised in Harper's Weekly, 20 January 1877 (which we will try answer one by one):

"What was the cause?"

We could assume that his question was answered. If the collapse of the Dee Bridge still to this day remains partly unanswered, the Ashtabula Bridge was having a rather unfortunate cast-iron detail (the lug) to transfer the load from the upper chord to the other truss members in the joints, and this lug broke in one place, enough to cause the complete collapse of the entire bridge.

"Was it improperly constructed?"

In the light of sustaining more than eleven years under service load it is hard to say that the bridge was a "scamped work" – the intentions were to build a strong, robust and safe structure – but still it had some rather strange and unfortunate characteristics (too much of the inspiration perhaps coming from the wooden Howe trusses). The jury finally stated in their conclusions: "Iron bridges were then in their infancy, and this one was an experiment which ought never to have been tried or trusted to span so broad and deep a chasm."

"Was the iron of inferior quality?"

It is not a matter of good or bad quality of the cast-iron material – cast iron is not a reliable material with respect to the safety of load-carrying structures, especially such subjected to tension (as cast iron is having very low ductility). Cast iron was very soon, after the collapse reasons of the Ashtabula Bridge became known, forbidden for future use in bridge structures.

"After eleven years of service, had it suddenly lost its strength?"

The understanding of fatigue and fracturing was not common knowledge among professional engineers during the 1800's...

"Or had a gradual weakness grown upon it unperceived?"

...but still there were some clever assumptions pointing in the right direction!

"Might that weakness have been discovered by frequent and proper examination?"

If one had known exactly what to look for, yes. And if the detailing had considered the need for close and thorough examination, yes. But now the bridge superstructure was "hidden" below the track, making access hard and limited. Visual inspection of many of the lugs was also impossible as they were hidden behind the surrounding I-girders (as the case was for the lug that broke), and looking for air holes inside the lugs would have required X-ray or ultrasonic apparatuses, devices which were not available in the 1800's.

"Or was the breakage the sudden effect of intense cold?"

The writer of the article had apparently an intuitive understanding of the mechanism and the parameters influencing the fracture toughness of steel.

"If so, why had it not happened before in yet more severe weather?"

This is quite a relevant question. The combination though, of a propagating fatigue crack that had reached a certain critical length (with respect to brittle fracture), given the temperature level, led to the initiation of unstable crack growth that particular night.

"Is there no method of making iron bridges of assured safety?"

Today it exists, but in those days it was very much (to say the least) up to the designer and the owner to decide upon the construction method, the material to use, the load levels, inspection routines and so on. The methods today of making safe and reliable bridges are to a large extent based on the experience of a number of failures in the past – it is from them that we have learnt.

"And who is responsible (so far as human responsibility goes) for such an accident – the engineer who designed the bridge, or the contractor, or the builders, or the railroad corporation?"

Of course the manufacturer of the bridge members and the contractor are responsible (of delivering safe and sound parts, and to build according to the drawings), but the railway company and the designer(s) are those that have the main responsibility.

"Was the bridge, when made, the best of its kind, or the cheapest of its kind?"

The best according to the railway company and the designer, but the cheapest with respect to the simple solutions regarding the truss members and detailing.

"Was the contract for building 'let the lowest bidder', or given to the most honest, thorough workmen?"

The contract was given to a carpenter (!), whose prior experience was limited to wooden bridges only. But given the fact that all-iron Howe trusses were scarce at that time it was not something astonishing.

"These and a hundred similar queries arise in every thoughtful mind, and an anxious community desire information and assurance of safety. The majority of people can not, of course, understand the detailed construction of bridges, but they do desire confidence in engineers, builders, contractors, manufacturers, who have to do with the making of them, and in the railroad companies, into whose hands they are constantly putting their own lives and the lives of those dearest to them."

These questions and this final remark really show that you do not need to be a qualified engineer to prove that you have a proper understanding of the key issues with respect to bridge design – something that was to a large extent neglected by Amasa Stone (and his brother). A new bridge was soon erected after the bridge failure, but this time they chose to build a wooden Howe truss.

Chapter 3

Tay Bridge

When the task of bridging the two great estuaries in Scotland – the Firth of Forth and the Firth of Tay – was commissioned to Thomas Bouch in the early 1870's, these had been giant barriers for efficient railway traffic in these parts of Great Britain since many decades (Fig. 3.1).

It was Thomas Bouch himself who had suggested the idea already back in 1849 when he was appointed engineer and manager of the North British Railways. The railway companies in the western part of Scotland were hard competitors carrying goods and passengers to the north of Scotland, and he knew that in order for the North British to survive they had to bridge the two estuaries. However, the idea was not supported by the NBR directors, dismissing it as "the most insane idea that could ever be propounded". Also the other railway lines did not approve of the idea, as

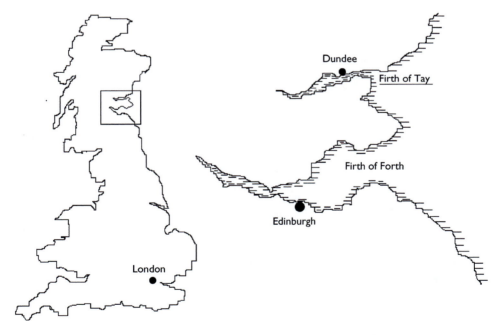

Fig. 3.1 Edinburgh (the capitol of Scotland) and Dundee are separated by two great estuaries – the Firth of Forth and the Firth of Tay.

Fig. 3.2 The Tay Bridge, which was built in 1878. (Wolcott: The breaks of progress. Mechanical Engineering)

it radically would change the position of power to the advantage of the NBR. In the mean time – when the opinion slowly came to support his idea – Bouch made the traffic more effective by introducing paddle-wheel train ferries over the two estuaries, but still the travelling between Dundee and Edinburgh was slow and troublesome. Finally, in 1871, the decision to bridge the Firth of Tay was taken, and two years later it was also decided to bridge the Firth of Forth. James Cox, a wealthy business man from Dundee, invested money in the Tay Bridge project – and by doing so had turned the opinion. He stated: "... that it would be for the public advantage, and tend greatly to the traffic of the north of Scotland and specially the town and trade of Dundee."

The construction of the Tay Bridge was originally planned to be completed in 1874, to a cost of little more than £200,000, but was delayed due to – among other reasons – inadequate sounding of the river bed (design modifications of the bridge substructure had to be made due to larger depth to solid rock than was expected). The bridge was finally opened in 1878, to a cost of more than 60% of the original estimate. As completed, the bridge became the longest railway bridge in the world (more than 3 km long) and it was regarded as an outstanding achievement (Figs. 3.2 and 3.3).

The bridge carried a single railway track, and in total the bridge consisted of 85 riveted wrought-iron truss spans, 72 shorter deck spans and 13 centrally located and elevated high-girder spans to allow for the navigation of ships. These high girders were so called trough-trusses, meaning that the trains passed in between the trusses in order to give maximum clearance in height for the ships in combination with a minimum

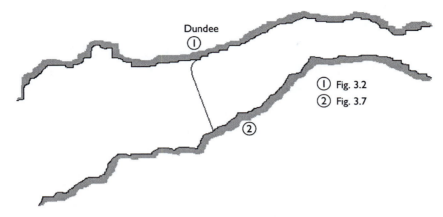

Fig. 3.3 The photo in figure 3.2 is taken from the north shore of the Firth of Tay (on the Dundee side).

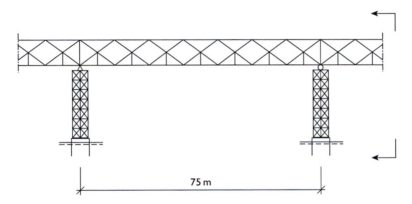

Fig. 3.4 The centrally located and elevated high-girder spans – a continuous system in sections of 4–5 trusses.

of ascent for the trains. Eleven of these elevated spans were 75 m and two were 69 m (Fig. 3.4).

The high-girder spans were supported by truss columns, having six cast-iron circular hollow profiles (380 and 460 millimetres in diameter) braced with wrought-iron diagonal and horizontal members (Fig. 3.5).

The through-trusses were braced at the top (as a frame) in order to ensure stability in the lateral direction against swaying – the clearance inside was still adequate for the trains. Another additional stiffening member could be seen in the elevation of the high girders (see Fig. 3.4), and which also was present in the deck girders. In order to avoid bending of the lower chord of the high girders – and the upper chord of the deck girders – a secondary vertical member was transferring the load from the transverse floor-beams of the deck to the truss. The load from a train is lifted up (through tension) to the joint of the diagonals in the former case (the high girders), and transferred down (by compression) in the latter (the deck girders). This small but

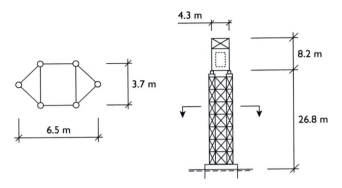

Fig. 3.5 The truss columns supporting the elevated high-girder spans.

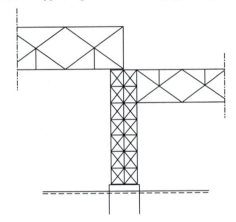

Fig. 3.6 Additional vertical members – from the lower chord of the high girders and from the upper chord of the deck-spans – help to avoid bending in the chords.

very important secondary member was completely left out in the Ashtabula Bridge (see Chapter 2) (Fig. 3.6).

Bouch had built railway bridges before – more or less similar in type to that of the Tay Bridge – but none had been as magnificent as this. He had described the project as "a very ordinary undertaking" when trying to convince the financiers, knowing that it would be quite the opposite, but now when it was completed he took great pride in all the acclaims he received.

On 20 June 1879 – one year after the opening of the bridge – Queen Victoria did cross the bridge on her way back from Balmoral Castle in the north of Scotland (the Royal summer residence), and described the passage over the Tay Bridge:

"We reached the Tay Bridge station at six. Immense crowds everywhere, flags waving in every direction, the whole population out; but one's heart was too sad for anything. The Provost, splendidly attired, presented an address. Ladies presented beautiful bouquets to Beatrice and me. The last time I was in Dundee was in September 1844, just after Affie's birth, when we landed there on our way to Blair, and Vicky, then not four years old, the only child with us was carried through the

crowd by old Renwick. We embarked there also on our way back. We stopped
here about five minutes, and then began going over the marvellous Tay Bridge,
which is rather more than a mile and a half long. It was begun in 1871. There
were great difficulties in laying the foundation, and some lives were lost. It was
finished in 1878. Mr. Bouch, who was presented at Dundee, was the engineer. It
took us, I should say, about eight minutes going over. The view was very fine."

The following week Bouch was invited down to Windsor where he was honoured
a knighthood from the Queen, and was from now on entitled Sir Thomas Bouch.
Everyone was not as enthusiastic though about the magnificence and safety of the
bridge. The Times wrote that "... when seen from the heights above Newport it looks
like a mere cable slung from shore to shore", quite rightly pointing out the rather
slender proportions of the bridge. John Fowler, a well reputed bridge engineer (and
the one who eventually became the designer – together with Benjamin Baker – of the
Forth Bridge), questioned the stability of the Tay Bridge and had therefore forbidden his family ever to cross over it (perhaps Fowler also had learnt about a certain
incident in February 1877, when two high girders were damaged in a storm during construction). How true his premonitions came to be. On the evening of the 28
December 1879 – 18 months after the grandiose opening of the bridge – a mail train
with six carriages was expected to arrive at Dundee Station at 07.15 P.M. A fierce
and violent gale had been blowing all day, but the traffic over the bridge had not
been stopped because of that. When the evening came the wind gusts had become
increasingly harder, and the people in Dundee was recommended to stay inside. As
the mail train, at a speed of approximately 30 km/h, finally entered the bridge on
the south side, it was signalled to Dundee Station that it was soon to arrive in just
7–8 minutes. The minutes went by, but no train arrived at Dundee. Attempts were
made to signal back to the south side, but the line was dead. It was finally decided
to send a man out on to the bridge, in the dark and stormy night, to see what had
happened. As he fought his way out into the darkness one could imagine the questions
going on in his mind: Had the train stopped on the bridge? Why was the telegraph line
out of order? Was the train held back because of the heavy storm? Was it something
wrong with the train? Was it still standing on the bridge, or had the driver put the
engine in reverse and turned back to the south bank? Had the train derailed? Or the
worst, but unthinkable scenario; had something happened to the bridge? As he eventually arrived where the high-girder spans were supposed to begin, there was nothing but
empty space – the bridge was just gone! Three years but one day after the Ashtabula
Bridge collapse in the USA, the Tay Bridge also collapses during the last days of December, and almost at the same time in the evening, and also during a violent storm – quite
an unusual (and sad) coincidence. The first telegram to reach the outside world read:

"TERRIFIC HURRICANE STOP APPALLING CATASTROPHE AT DUNDEE
STOP TAY BRIDGE DOWN STOP PASSENGER TRAIN HURLED INTO
RIVER STOP SUPPOSED LOSS OF 200 LIVES STOP"

The following morning, when the storm had calmed down, the extent of the tragedy
became evidently clear. The entire high-girder section had collapsed into the river; close
to one kilometre (!) of the bridge gone (Fig. 3.7).
    The train – and its passengers (including Bouch's son-in-law) – had gone down with
the bridge, taking 75 lives (no one on the train survived). There is no doubt that the

Fig. 3.7 All thirteen high girders of the Tay Bridge had collapsed into the river. (Wolcott: The breaks of progress. Mechanical Engineering)

storm was a major contributing factor why the bridge had collapsed, but then, should not bridges be able to withstand that kind of loading (still being built during the 1800's). And was it just the wind itself that had caused the collapse, or was it something more? A lot of things could be said about the causes of the collapse (and so I will), but first and foremost it must be made clear that this gigantic project was under a lot of pressure due to the delays and a tight financial budget, and this unfortunately led to anything but a meticulous attitude towards manufacturing and assembly, but above all it was a matter of a total negligence in design towards proper care regarding wind loading. Bouch admitted to the Inquiry that the combined effect of wind and traffic had not been considered in the design, and it was because of this that no extra horizontal wind bracing had been added to the bridge below the deck. Before the start of the project he had consulted the Chief Inspector of the Railways, Colonel William Yolland, with the question; "is it necessary to take the pressure of wind into account for spans not exceeding 200 feet span, the girders being open lattice work?", and was given support from Yolland in this view. Perhaps the permeability of the Ironbridge, that easily withstood an extreme spring flood in 1795 (see Chapter 1) – in contrast to the solid walls of the stone arch bridges along the river Severn – had the engineers to believe that open lattice structures (i.e. trusses) were not resisting wind (as well as water for that matter). Wind as a design parameter was in fact something that not only Bouch was regarding as a minor problem – it was generally considered that the combined loading from self-weight of the bridge and the weight of the traffic (together with the horizontal thrusts from the trains in the transverse direction due to out-of-alignment of the track) was sufficient with respect to the required load-carrying capacity and stability of the bridge. At that time there existed not any up-to-date wind pressure values in Great Britain to be used in design. The values that were given dated back to the mid 1700's, and were, for example, less than one fourth of the values used in France and the USA; however, it was the designer's responsibility to provide for the prevailing wind in each project, and given the fact that this was a high elevated structure over open water (and a wide passage), exposed for very strong gales, it must be concluded that Bouch was negligent (to say the least) in this case. And as the wind pressure values given were low, he must have – quite simply – regarded this effect on the bridge's stability to be negligible, and the combined loading of wind and traffic not being a loading case to consider. The question to Yolland also revealed that the focus with respect to wind was on the structure itself, rather than on the rolling load – perhaps wind pressure on the surface of the train was regarded as a highly temporary and negligible effect.

As we have learnt the bridge was already from the beginning questioned with respect to stability and safety, but also after the stipulated load test (before taking the bridge into service) it was decided by the Inspector of the Board of Trade, General Hutchinson, that there should be a speed limit of 25 mph (40 km/h) for the trains. This measure was most certainly first of all because of the slope of the track and the safety against derailment, but could also have been because of doubts regarding the stability of the bridge in high winds.

Already from the start the bridge showed signs of distress and the evidence of poor quality material. There were reports about the cast-iron circular hollow profiles being defective (containing longitudinal slits and hidden repairs of large blowholes), cracks in the brick piers and of scour of the river bed. But more alarmingly, the bracing diagonals (the ties) in the truss columns (i.e. the towers) had in many locations worked loose shortly after the bridge was taken into service. A re-painting crew reported about a rattling noise coming from the bracing diagonals when trains were passing overhead, as well as a side-to-side swaying (oscillation) of the truss columns. Passengers in the trains also complained about strange vibrations. Taken together, these observations were clearly indicating that something was wrong.

Concentrating on the truss columns in the continued discussion regarding the governing weakness of the bridge, we start with a brief description of the joint configuration. In each joint the wrought-iron horizontals and diagonals were attached (bolted) to the vertical cast-iron cylinders (the circular hollow profiles) through cast-iron gusset plates (lugs), which were cast in the same mould as the cylinder, as integrated parts (Fig. 3.8).

In order to allow for an easy erection of the truss members – especially the diagonals – one end was provided with an elongated hole (to allow for tolerances due to slight differences in length). At assembly the diagonals were fastened to the gusset plates, and then secured and tightened through small tapered cotters ("wedges") which were fitted in the over-lapping gap. It was these cotters that for some diagonals had worked loose (through the shakings and vibrations), making the function of these truss members heavily impaired – now more or less comparable to a non-active zero bar. But there were other circumstances that made this truss column weaker than perhaps was the intention. By choosing flat bars the diagonals of the truss column could not transfer compressive load due to the negligible buckling strength of these members. So here we had a structure comprising of diagonal elements:

- that easily lost their strength in tension, and
- that were more or less unable to carry any load in compression.

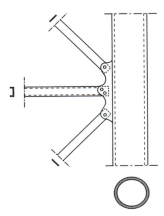

Fig. 3.8 The configuration of the truss column joints. The diagonals were simple flat bars and the horizontal was a U-channel bar (possibly a T-bar).

At a first glance the truss column has the appearance of an inner statically indeterminate structure, but which in fact is statically *determinate* due to the inactive compressive diagonals, and could be transformed into a "mechanism" with respect to the action of a true truss (when diagonals in tension also are lost). In comparison to the Ashtabula Bridge – where all members (in tension and in compression) were designed with respect to maximum stress, rather than taking the risk of buckling into account for the members in compression – the Tay Bridge design led to a utilization of the diagonals in tension to twice the intended value. In a truss having equally stiff crossing diagonals, the transfer of load is divided equally in between the diagonals. However, if we study the transfer of a horizontal load in a vertical cantilevering truss – having one of the diagonals in each panel unable to carry load in compression – the resulting action is that the load is transferred down into the base as a constant force (in tension) in only one set of the diagonals (Fig. 3.9).

The result is that the active diagonals (these in tension) are carrying twice the load, i.e. at a double strain. The vertical members though – in contrast to the horizontals and the diagonals, where the force is constant over the entire length – have a gradual increase in the normal force down into the base, and are not dependent upon whether one or two diagonals are active in each panel.

So, we have a situation where the diagonals in compression are inactive, and the diagonals in tension have either worked loose (due to the loss of the cotters) or are about to loose their strength due to excessive loading. The cast iron gusset plates (to which the diagonals are attached) are not able to carry any heavier loading – being of a brittle material as they were – so these could fracture very easily as well. The cast iron material has little capacity for plastic deformations (enabling for the redistribution of load) and will rupture in a brittle manner without any prior yielding.

We study the set of panels closest to the base of the truss column (where the force in the diagonals are slightly higher than the panels above, given that we take into account the wind loading on the truss column itself) (Fig. 3.10).

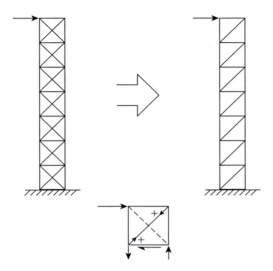

Fig. 3.9 The resulting transfer of "shear" in a vertical cantilevering truss if one diagonal of each pair in the panels is unable to carry compression (because of a negligible buckling strength).

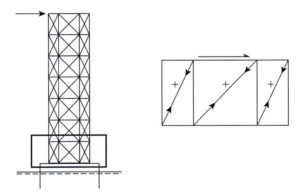

Fig. 3.10 The resulting action of the bottom panels in the truss column (as well as every set of panels), considering that the diagonals in compression are inactive.

The major part of the horizontal load is transferred down through tension in the active diagonal of the inner panel, and if this diagonal is lost – having worked loose due to the loss of the cotters, or due to local failure of the connection – then all load has to be taken by the remaining two diagonals in the outside panels.

As the remaining two diagonals also are lost – due to the extreme loading – then these bottom panels have been transformed into a local frame (Fig. 3.11).

And working locally as a frame, the truss column cylinders are now not only subjected to normal forces alone, but also to bending moments. Bending in the locations where the gusset plates are attached to the cylinders produce extremely high stress levels (due to the local stress concentration effects), so there is also a risk for an unstable

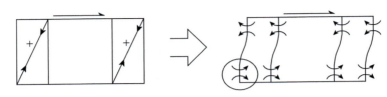

Fig. 3.11 The gradual loss of active diagonals (in tension) transforms the truss panels at the base into a frame. This local failure scenario is also very probable for the panels higher up, because of the vibration and the swaying here (making the diagonals more easily to work loose).

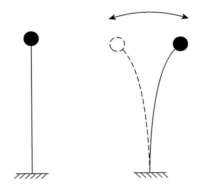

Fig. 3.12 A model of the Tay Bridge as a top-heavy cantilevering structure, which for sure it was.

crack to rupture the entire cylinder. We are now very close to the actual collapse of the bridge, but we will first study the truss column to see if there was any initiation of resonant vibration coming from the wind gusts. The static wind flow – which is the dominant force – is constant in time, and does not produce any vibration of a structure (such as a cantilever), just a static deflection. It is the variation of the pressure (read: the wind gusts) that could initiate resonant vibration, given that the frequency of these variations coincide with the natural frequency (eigenfrequency) of the structure itself, and then a dynamic amplification of the deflection will take place, which could – if the damping is low – produce forces in the structure which are very much higher than that of the static wind pressure. It is generally accepted that if the natural frequency of a structure (subjected to wind) exceeds 3–4 Hz, then there is no probability for resonant vibration coming from wind gusts (read: the frequency of wind gusts does not exceed 3–4 Hz, normally not even 1 Hz, at least not with any substantial energy to speak of).

Let us consider the weight of the bridge spans and that of the train as a concentrated mass on top of the cantilevering truss column (Fig. 3.12).

In order to determine the natural frequency of the truss columns for the high-girder bridge section of the bridge, we need first of all to find the second moment of area of the cross-section around the y-axis, as the deflection for the wind is along the x-axis (Fig. 3.13).

The thickness of these cylinders must be assumed, and the expected values to choose are 25 mm for the large cylinders and 20 mm for the small, giving the cross-sectional

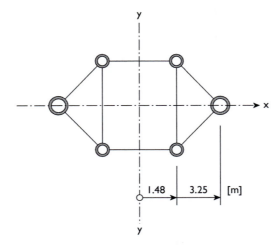

Fig. 3.13 The cross-section of the cantilevering truss column (see also Fig. 3.5). As a qualified guess the two exterior cylinders are those having the diameter of 460 mm, and the four interior having the diameter of 380 mm.

areas (Eqs. 3.1 and 3.2).

$$A_{\text{large}} = \pi \cdot 460 \cdot 25 = 36,128 \, \text{mm}^2 \tag{3.1}$$

$$A_{\text{small}} = \pi \cdot 380 \cdot 20 = 23,876 \, \text{mm}^2 \tag{3.2}$$

The second moment of area (reduced to 70% taking the "shear" deformations into account, i.e. the axial deformation of the horizontals and the diagonals) (Eq. 3.3):

$$I_{y-y} = 0.7 \cdot (2 \cdot (2 \cdot 23,876 \cdot 1480^2 + 36,128 \cdot 3250^2)) = 6.807 \cdot 10^{11} \, \text{mm}^4 \tag{3.3}$$

We also need the value of the concentrated mass on top of the cantilevering column, as the weight of the bridge span plus that of the train (taken from D.R.H. Jones) (Eq. 3.4):

$$M = 234 + 122 = 356 \, \text{tons} \tag{3.4}$$

The natural frequency can now be found (taking 100,000 MPa as the modulus of elasticity for cast iron, and the length of the truss column as 23.7 m) (Eq. 3.5):

$$f = \frac{1}{2\pi} \cdot \sqrt{\frac{K}{M}} = \frac{1}{2\pi} \sqrt{\frac{3 \cdot EI}{M \cdot L^3}} = \frac{1}{2\pi} \cdot \sqrt{\frac{3 \cdot 1 \cdot 10^{11} \cdot 6.807 \cdot 10^{-1}}{356 \cdot 10^3 \cdot 23.7^3}} = 1.04 \, \text{Hz} \tag{3.5}$$

There are, however, a number of influencing factors that we have not considered:

- The cylinders were filled with concrete, and this increases the axial stiffness of the same, but it also increases the weight.
- We have not included the mass of the truss columns.

- The decrease of the stiffness from diagonals working loose.
- The stiffening effect coming from the continuous high girders.

Taken together, these parameters have not a major influence on the result (as some would increase the result and some would decrease the same), and this analysis is not an exact calculation of the natural frequency, rather a "probability study" showing that there, indeed, was a risk of resonant vibration of the truss columns – due to wind gusts – when the mail train was passing the bridge.

In addition, there was also an incident during the construction of the bridge that very well could have contributed to the swaying of the bridge also during calm weather. A high girder section was damaged (slightly bent) during the lifting procedure (it was dropped into the sea), and yet this section was used. The drivers reported of a curvature in the track that produced a horizontal, transverse thrust each time the engine passed that particular section. Each axle of the train also contributed to this low-frequency excitation, and given the energy of these thrusts (in relation to wind gusts), perhaps this was a much more worse loading case for the bridge to handle. Resonant vibration could have been the case for many train passages (a function of speed in combination with the distance between the axles) and would in such case explain the complaints from the passengers. Resonant vibration or not, here was a structure subjected to side-ways swaying each time a train was passing and this could explain why the truss columns had many diagonals working loose. A gradual loss of stiffness lowered the natural frequency and made the structure more and more susceptible to these horizontal thrusts, and the combination of horizontal thrusts and wind gusts at that fatal evening made the entire high-girder bridge section collapse into the water. From the start, when the bridge was taken into service, the structure was very weak, and it became gradually weaker, so it was just a matter months (18 to be exact) before it would fail.

The wind was blowing very hard on the evening of the 28 December 1879 as the mail train approached the Tay Bridge from the south. The structure itself was of course subjected to the wind pressure, but the incomparably largest wind loading came when the train entered the bridge, and as it climbed towards the high-girder section the structure was strained to the utmost. Twenty-seven metres high up in the air – as a huge elevated signboard exposed to wind – there was pressure on the windward side of train as well as suction on the leeward side (Fig. 3.14).

The maximum pressure came from the wind gusts, exerting not only a maximum deflection of the truss columns in the wind direction, but also, most probably initiating resonant vibration (swaying) which amplified the forces – at the base of the truss columns – and the deflections of an already weakened structure, which finally became too much for the bridge to carry (Fig. 3.15).

Ten out of twelve truss columns fractured at the base (to the piers) due to a sudden loss of anchorage (one after the other) and the remaining two fractured some few metres above the base. As the diagonals had lost their strength – during the eighteen months the bridge had been in service and during the final stages of the progressive collapse – the loading at the base of the cylinders changed from being pure normal forces to the combination of normal force plus bending moment, neglecting the shear force being present. Here only the action coming from the horizontal wind load is shown (not considering the vertical compression coming from the weight of the train) (Fig. 3.16, see also Fig. 3.11).

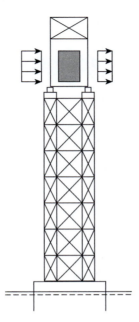

Fig. 3.14 The wind loading on the train twenty-seven metres high up in the air.

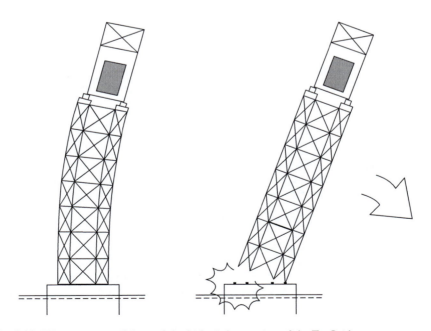

Fig. 3.15 The progressive failure of the high-girder section of the Tay Bridge.

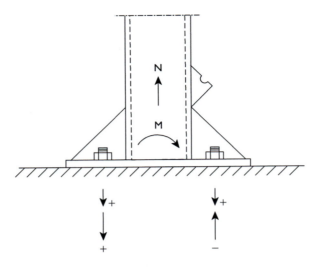

Fig. 3.16 As the diagonals were lost the loading at the base of the cylinders increased dramatically. Also shown above is the fractured gusset plate where the diagonal had been connected.

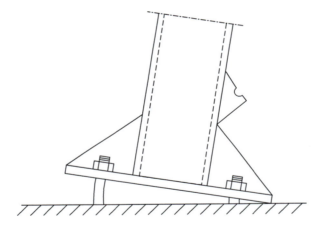

Fig. 3.17 Elongated and deformed anchorage bolt on the tension side of the cylinder base plate.

As a consequence the anchorage bolt at the tension side became overstressed (Fig. 3.17).

There were three possible major fracture scenarios at the base of the cylinder:

– A complete rupture (read: brittle fracture) of the cylinder above the foot, starting from the corner of an attached gusset plate on the tension side.
– Rupture of the anchoring bolts, starting with the most strained on the tension side.
– Loss of anchorage of the bolts to the brick pier.

An example of the latter – followed by the first fracture scenario – is shown in figure 3.18.

Fig. 3.18 One of the broken piers of the collapsed high-girder bridge section (looking north). (http://taybridgedisaster.co.uk/)

Afterwards, when the piers were examined, they found a large number of nuts and bolts together with broken-off pieces from the gusset plates (the outside parts after fracture of the net-section through the bolt holes – the missing parts of the gusset plates in Figs. 3.16 and 3.17).

The Court of Inquiry concluded that "the fall of the bridge was occasioned by the insufficiency of the cross bracing and its fastenings to sustain the force of the gale", and indeed so it was. The high-girder section of the bridge would definitively have benefited from an extra pair of straddling legs (like the intuitive position of a human body expecting a horizontal thrust), stabilizing the truss columns and drastically lowering the strain due to severe winds (Fig. 3.19).

For the deck girder section in the curved part of the bridge Bouch had chosen to stabilize the truss columns against the centrifugal forces having a single-sided straddling leg (see Fig. 3.2), but for the straight part of the bridge this measure was judged superfluous, which in hindsight was a pity. As we have learnt, the horizontal forces coming from the wind and the side-ways thrusts of the trains were not regarded by Bouch to be any problem (which, in fact, actually was the case for the lower part of the bridge, as it could withstand the gale), but would he have considered to stabilize the truss columns – especially of the high-girder section – then there would have been a need for much wider piers, which would have delayed the project even more and increased the cost.

The negligence of not taking proper care of the wind pressure (the magnitude as such, but also of how wind is affecting high elevated structures carrying traffic) – not only by Bouch in the Tay Bridge project, but also by the engineering profession at large – had the immediate effect that the wind forces to be used in future designs were

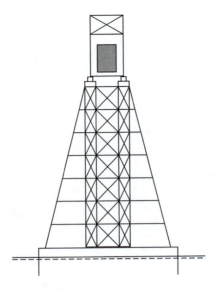

Fig. 3.19 An extra pair of straddling legs would most probably have saved the Tay Bridge.

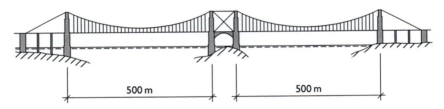

Fig. 3.20 The project by Bouch to span the Firth of Forth by a suspension bridge (principle lay out).

drastically increased. Another immediate result of the Tay bridge disaster was that the ongoing project to bridge the Firth of Forth was stopped. Bouch was constructing a huge suspension bridge, which would become the largest in the world, having two main spans of about 500 metres each (Fig. 3.20).

The longest suspension bridge in the world to carry railway traffic at that time was the Niagara Falls Suspension Bridge (by J.A. Roebling), being only 250 metres, but still it was stabilized with a large number of guy wires to counteract for the swaying effects from the trains and from the winds. Given that the proposed Forth Bridge would have been twice as long as the Niagara Falls Bridge it really shows that Bouch was in over his head with respect to realistic projects. Less than a year after the collapse Sir Thomas Bouch died.

In 1887, eight years after the collapse of the Tay Bridge, a new double-track bridge was erected parallel to the old one, and the wrought-iron girders from the old bridge (from the deck-girder spans) were re-used. In contrast to the old Tay Bridge the

Fig. 3.21 The new railway bridge over the Firth of Tay, built in 1887, having the same number of high-girder spans as the old bridge. (www.mccrow.org.uk/TaysideToday/TayBridges/ TayBridges.htm, with kind permission of Malcolm McCrow, photographer and copyright holder)

new bridge had very strong and stable intermediate supports – wrought-iron tubular arches – not only coming from the wind loading requirements, but also because of the increased width of the track superstructure (Fig. 3.21).

To the right of the new bridge the brick piers of the old bridge are seen (see the figure above). They were most probably saved because of two reasons, besides acting as breakwaters – first of all that the removal of the same was judged to be too much of an extensive project, but also because that they would function as good protection against ship collisions (at least from one direction). Spectacular bridge failures linger on in people's minds sometimes for ages, but the failure of the old Tay Bridge is especially vivid as the old brick piers were kept.

The deck girders from the old Tay Bridge were not the only parts that were re-used, but also the engine of the mail train was taken into service again, as it was salvaged from the wreckage of the bridge and repaired (Fig. 3.22).

Three years after the completion of the New Tay Bridge, the Forth Bridge was completed in 1890, still in use today (as the Tay Bridge also is) and being one of the most impressive bridges in the world (Fig. 3.23).

The lessons learnt from the Tay Bridge disaster were adapted to the full in the construction and design of both the New Tay Bridge and the Forth Bridge. Maximum stability was ensured and the materials used were tested meticulously – nothing was left to chance. In the case of the Forth Bridge, such an extraordinary design would also not have existed if it was not because of the Tay Bridge failure (as a consequence the cost to build the Forth Bridge was nearly almost ten times higher than that of the old Tay Bridge). The use of cast iron was completely abandoned, and steel was for the first time introduced in bridge design (to the full in the case of the Forth Bridge). In a way

Fig. 3.22 The engine that was salvaged from the collapse of the Tay Bridge – "The Diver" as it was called. (Yapp, N.: 150 Years of Photo Journalism)

Fig. 3.23 The huge cantilever truss bridge over the Firth of Forth in Scotland, which replaced the suspension bridge proposed by Bouch. This bridge surpassed the Brooklyn Suspension Bridge from 1883 (486 m) as the longest span in the world (521 m).

Thomas Bouch had had the good judgment of choosing cast iron for the elements in dominating compression (the cylinders), and wrought iron for the members subjected to tension (the girders and the cross-bracing of the truss columns), but he was forced to accept cast-iron gusset plates given the choice of cylinder material. Had also the cylinders (and consequently the gusset plates) been of wrought iron, then the strength and toughness of the structure perhaps would have been able to withstand the strain, but still the structure would have been swaying – susceptible to wind as it was – and eventually other problems would have occurred, such as fatigue. Bouch had not experienced any stability problems in his earlier bridges of similar design, and that was because that these bridges – as well as the deck girder spans of the Tay Bridge – were not as elevated as the central high-girder spans. The key parameter here is the height, and this – together with the unfortunate choice of cast iron material for the gusset plates (read: for the cylinders) – made the bridge eventually collapse. A faulty structure such as the Tay Bridge was doomed already from the beginning it was taken into service, and it was the wind that for sure pushed it over the edge. And even if the traffic would have been stopped that evening – like what is done today for exposed bridges in extreme weather – the Tay Bridge was on the verge of collapsing, and would have done so sooner or later for more or less any given wind.

Chapter 4

Quebec Bridge

During the 1800's the St. Lawrence River – stretching as it is more than 1000 kilometres from Lake Ontario (in the Great Lakes Region) to the Atlantic Ocean – was a great barrier for the growing commerce in the south-east corner of Canada (Fig. 4.1).

During the first half of the 1800's the City of Quebec had the advantage – to the other major city in the French-speaking province of Quebec, i.e. Montreal – that ocean-going sailing ships could not go any further than to Quebec (City). This advantage was partly lost when steamships were introduced, as cargo now could be transported all the way up-stream to Montreal. When Montreal, in the mid-1800's, decided to build a bridge across the river, Quebec was even more falling behind the competition.

The Victoria Bridge at Montreal (named after Queen Victoria) was built in 1859, and was the first bridge to span the wide expanse of the St. Lawrence River. The designer was Robert Stephenson from England, son of the famous railway engineer George Stephenson. Robert Stephenson had constructed the monumental Britannia Bridge over the Menai Straits in Wales in 1850, and he used the same design for the Victoria Bridge, namely a closed rectangular box-girder section, having a maximum span length of 100 metres. The total length was nearly 2 kilometres, making it the longest bridge in the world (however, to be surpassed by the Tay Bridge in 1878). The need for an increased load-carrying capacity, together with complaints from the passengers about the smoke from the locomotive, made it necessary to rebuild the bridge in 1897–1898 – the tube walls were then replaced by open trusses.

Quebec City also needed a bridge, not only to compete with Montreal, but also to expand to the south side of the river, and to open up for railway transports. In the late 1890s, the Quebec Bridge Company was formed, consisting of political leaders and businessmen. Despite a shortage of funding the project was launched, and a suitable site where the bridge was to cross the river was chosen (just some few kilometres downstream of the City of Quebec). Government funding was given, but still the project was under a tight financial constraint, and remained so for the entire design and construction process.

Perhaps in an attempt to boost the project and to raise funds, the renowned bridge designer Theodore Cooper was engaged in the project in 1897 (at the age of almost 60). Cooper had had a long carrier in constructing bridges, starting in 1874 with the famous Eads Bridge over the Mississippi in St. Louis. James Eads, the chief designer of the bridge, had appointed Cooper in charge of the erection of the bridge – an arch bridge in steel, the longest in the world of that type (having a main span of 155 m). This bridge was the first major bridge in the world to use steel for its main components

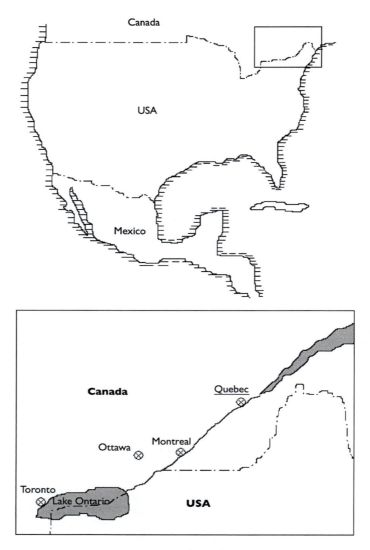

Fig. 4.1 The St. Lawrence River in Canada from Lake Ontario to the Gulf of St. Lawrence, and the major cities in this area, which were cut off from the Canadian ports at the Atlantic Sea in the east during the 1800's. See also figure. 2.1.

(eight years earlier, in 1866, a short railway bridge of some 40 metres was built over the Göta Älv in Vargön , Sweden, near the city of Vänersborg – perhaps the first ever steel bridge in the world). Steel as an engineering material had (and still has) a carbon content in between that of wrought iron and cast iron, giving both high compressive and tensile strength, together with adequate ductility (i.e. an ability to plastify without fracturing at excessive loading). The material was at that time new and unproven, but was gradually becoming the dominating material in bridge construction – being the obvious choice for the Forth Bridge and the New Tay Bridge, and of course also for the Quebec Bridge which was to be built.

Besides constructing many bridges, Cooper had also contributed by publishing many important papers on bridge design, and this made him the perfect choice for the Quebec Bridge Company in guaranteeing the project success. He was also known for giving all his projects complete and full attention (down to the smallest detail). However, Cooper was not to be the actual designer of the Quebec Bridge; instead his expertise and knowledge should ensure that the contract was given to the best and most competitive proposal. He was from then on to act as a chief designer, supervising the actual construction process (i.e. the design, fabrication and erection of the bridge). As the cost was a major financial constraint the lowest bid stood a good chance of winning the contract. Cooper himself had agreed upon half his normal fee in order to cut costs, but also to make it possible to be part in a final and major project crowning his career.

Among the many proposals that Cooper had to consider were suggestions for different suspension bridges, but he favoured a proposal from the Phoenix Bridge Company in Phoenixville, PA, just outside Philadelphia (see Fig. 2.1). He found that the proposal from the Phoenix Bridge Company was the "best and cheapest". Phoenix Bridge was a well-established company with a good reputation in producing mainly railway bridges of moderate sizes and length, but standardized so that production became tailor-made. They had suggested a riveted cantilever truss bridge for the Quebec Bridge (having some experience with that particular bridge type in at least one earlier bridge project, the Red Rock Bridge in 1890), which also was the best alternative according to the French engineer Gustave Eiffel (not only the designer of the famous Eiffel Tower, but also to many impressive railway bridges). Earlier in this process, Quebec Bridge Company had also considered a suspension bridge (hiring the famous bridge engineer Gustav Lindenthal to do the preliminary plans) but this idea was dismissed, probably due to the problems related to the flexibility (i.e. lack of sufficient stiffness) of these kinds of bridges – a cantilever truss bridge is very much more rigid and stiff, making it more suitable for train loading. An arch bridge was also not an option (although being a very stiff bridge type), as the needed span length was very long in combination with the requirements during erection that under no circumstances should the traffic on the river be obstructed (but still Cooper had used the cantilever erection method for the Eads Bridge – because of the same restrictions – but then for a much shorter span, 155 metres).

The cantilever truss bridge suggested by the Quebec Bridge Company, consisted of three truss spans (and two smaller approach spans), having a main span of 549 metres over the river (Fig. 4.2).

Originally, the main span of the bridge was planned to be 488 metres (1600 ft), but Cooper had proposed an increase of the same to 549 metres (1800 ft), by moving the piers closer to the shore (less problems with ice floes and easier to build, thus saving costs). Most certainly he also saw the chance to break the world record held by the Forth Bridge in Scotland (see Fig. 3.23), being 521 metres (1710 ft). 1800 ft was a nice and round figure, exceeding the Forth Bridge by 90 ft, and being the new world record it would even more draw the attention not only to the project, but also to Cooper himself. He also saw an opportunity to beat the Forth Bridge with respect

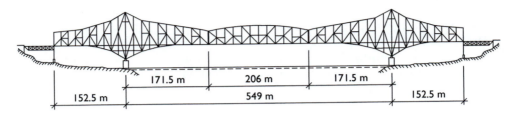

Fig. 4.2 The proposed cantilever truss bridge over the St. Lawrence River outside Quebec, similar in type to that of the Forth Bridge in Scotland.

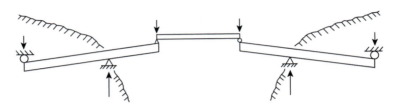

Fig. 4.3 The cantilever girder bridge using timber logs to bridge over a deep ravine.

to the amount of material used, and by doing so save as much money as possible for the project. Cooper commented upon the Forth Bridge in these words:

> "The clumsiest structure ever designed by man; the most awkward piece of engineering in my opinion that was ever constructed. An American would have taken that bridge with the amount of money appropriated and would have turned back 50% to the owners instead of collecting when the bridge was done, nearly 40% in excess of the estimate."

Many people of that time (i.e. not only Cooper) regarded the Forth Bridge of being too "bulky", but they did not fully understand the pressure on the designers in the aftermath of the Tay Bridge disaster – a safe and more or less excessively strong bridge had to be delivered, taking absolutely no chances whatsoever. However, Cooper, over-confident as he was, saw his chance here – together with the Phoenix Bridge Company – to deliver, not only the longest bridge in the world, but also perhaps the most optimal, using as little material as possible in order to save costs. Similar to the Forth Bridge, the Quebec Bridge was also to carry a double railway track, but in addition room was also to be made for trams and cars, making the bridge very wide. However, with respect to the design load, there is no difference between the Forth Bridge and the Quebec Bridge, as the trains are the dominant load.

The cantilever girder system is perhaps one of the oldest types used to bridge over wide waters and deep ravines in ancient time, where a simple log would not reach to the other end. By cantilevering out a log (or several logs) from both ends, and letting these two arms carry a mid-span log – simply supported – the maximum possible span length was drastically increased (Fig. 4.3).

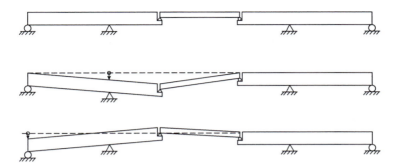

Fig. 4.4 Uneven settlements of the supports do not introduce bending in a Gerber beam; the members are inclined but remain straight.

The cantilever arm is embedded in the ground, counterbalancing the mid-span girder by the weight of earth and rocks, which is holding it down. The cantilever girder bridge using logs was perhaps not that common in the Western World, but more frequently used in China and Tibet. However, as continuous girders of iron and steel was introduced in the West – mainly for railway bridges during the second half of the 1800's – a German bridge engineer by the name of Heinrich Gerber patented an ingenious system of his (called the "Gerber beam") using hinges in the zero-moment locations (see Fig. 4.5). By doing so, the statically indeterminate system was transformed into a statically *determinate*, but still being equally as strong. But the main advantage was its capability to adjust with respect to settlements of the supports – for any given combination of uneven settlements among the supports the girders remain straight, therefore not introducing additional secondary stresses which could endanger the safety of the bridge (Fig. 4.4).

The Gerber beam system is also advantageous with respect to temperature deformations (elongations). The only disadvantage though – because of the system's flexibility – could be if *stiffness* is a design parameter. A concentrated load at the end of the cantilever arm is only taken by this arm (and is not distributed to the other arm), which could give a need for deeper girders in order to reduce the deflections.

The first modern bridge of this kind was built by Gerber over the River Main in Germany in 1867 – a three-span girder bridge with a centre span of 38 metres.

To transform a statically indeterminate continuous *truss* into a "Gerber system" was also very advantageous in those days, as graphical methods for determinate trusses such as the Cremona diagram or the Culmann diagram existed, making it relatively easy to analyse the member forces by simple static equilibrium in each joint.

For a continuous girder – having constant depth – the strain will vary according to the magnitude of the bending moment (large moments – large stresses, small moments – small stresses). However, if the girder depth is varied according to the moment distribution the normal stress could be held approximately constant for the girder (the same applies for the member forces of the upper and lower chords in a truss). Consider the moment distribution (for evenly distributed load) of a continuous girder (Fig. 4.5).

M:

Fig. 4.5 The moment distribution of a continuous girder (or Gerber beam) for self-weight and evenly distributed load. The highest values are found in the negative bending moment region over the two inner supports.

It is easy to see how the cantilever truss bridge is influenced by this moment distribution (see also Figs. 4.2 and 3.23):

–   Hinges where the moment is zero,
–   Increased depth over the inner supports in order to keep the stresses down,
–   Parabolic shape for the suspended span (however, turned upside down to give maximum clearance for the ships on the river),
–   An overall shape which, by following the bending moment distribution, gives more or less constant normal force in the upper and lower chords, making the cross-sections nearly the same (simplifies production and assembly),
–   A shape which in addition is "lifted up" in order to give necessary headroom for the railway trains,
–   A reversed (i.e. mirrored) shape over the inner supports in order to further reduce the normal forces in the chords, and – what is more important (especially for a Gerber system) – to increase the stiffness of the cantilever arm. And of course also for symmetry and aesthetics.

In 1900 the work on the substructure (foundation, piers and abutments) began, but it was not until four years later (i.e. in 1904) that the actual work on the steel superstructure could begin. Lack of funding had slowed the process down to a minimum during the three-year period between 1900 and 1903, and it was first when the Canadian Government guaranteed financing that the assembly of the superstructure could commence. The steel work began on the south shore of the river by the construction of the 152.5 m long anchor arm on falsework (i.e. temporary scaffolding) (Fig. 4.6).

As soon as the south anchor arm was completed (in early 1906) the assembly of the cantilever arm could begin (the falsework was then also moved to the other side of the river for the construction of the north anchor arm). Using the anchor arm as a counterweight – securely anchored at the end for the uplift forces (see Fig. 4.3) – the main span was being erected using the cantilever technique (one half from either side of the river), the same construction sequence that had been used for the Forth Bridge in 1889. And getting the inspiration from the human model so often shown at the construction of the Forth Bridge, one could similarly, for the Quebec bridge, exemplify the flow of forces and equilibrium of the anchor arm balancing the cantilever arm at this stage of the construction process (Fig. 4.7).

The free cantilevering technique – made possible by the continuous system – had the advantage that it combined the need for the passage of ships during construction, together with a free and simple assembly over open water.

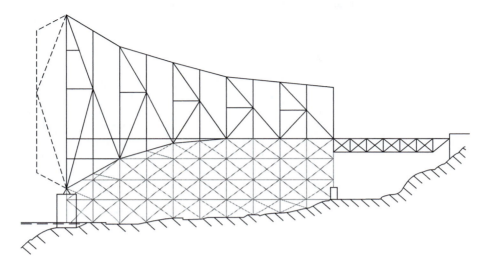

Fig. 4.6 The south anchor arm was erected on falsework.

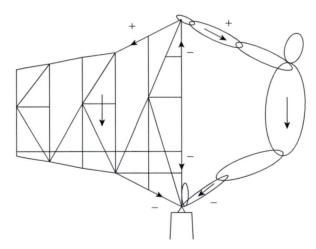

Fig. 4.7 The cantilever arm is balanced by the counterweight of an imaginary "being", pulling with the arms and pushing with the legs.

At just about this point in the construction (in early 1906) it was discovered that the amount of steel material delivered to the bridge site considerably exceeded the originally estimated amount – very strange to all those involved in the project, including Cooper. However, it was soon found that the self-weight used in the estimations was based on the early calculations made for the original span length of 488 metres. No correction had been made when the span length was increased from 488 m to 549 m. Every truss member had been adjusted with respect to the increase in span length, but not taking the increased weight into account – the member forces were consequently based on a span length of 549 m, but having a self-weight loading coming from the shorter span length of 488 m. Cooper judged this error in design to be within the limits of the load-carrying capacity – he calculated the increase in stresses to be around 10%, and concluded that the safety margin was acceptable. However, this was a decision that was quite remarkable, as Cooper early in the design process already had allowed for extraordinarily high working stresses – very much higher than otherwise was common in bridge design in those days. Most likely Cooper stipulated that high allowable stresses were to be used in order to reduce the amount of material needed – he was so determined to prove his statement about the Forth Bridge bulkiness (in combination with the need of keeping the cost down of his own bridge of course). The maximum stress level used was 145 MPa for normal loading (165 MPa for extreme loading), and it is very high even with respect to steel codes of later date. If we compare Cooper's values with old steel codes in Sweden, we see that it is not until 1970 that the levels are the same for mild steel (yield strength of 220 MPa):

"Normalbestämmelser för järnkonstruktioner till byggnadsverk" (1931):

Normal loading:	115 MPa
Extreme loading:	140 MPa

"Stålbyggnadsnorm 70, StBK-N1" (1970):

Normal loading:	147 MPa
Extreme loading:	169 MPa

Of course it is a matter of design philosophy (safety factors and load combinations) in combination with the steel material as such (scatter in strength data from tests and the mean value of the yield stress level), but it is highly unlikely that steel in the beginning of the 1990's would be of higher quality than later on – Cooper for sure wrote his own standards.

The project was well on its way, and to reconsider the design was unthinkable – Cooper had also reassured that the load-carrying capacity was adequate, so everything continued as planned. Perhaps Cooper realized though that he was slowly losing his grip on the project, because already in 1904 he had asked to be relieved of his commitments due to, as he said, his bad health (he was then 65 years old), but was persuaded to continue. He was now taking part in a project where cost was more important than safety, and where the normal forces of the truss members in the steel structure were not only higher than in other comparable bridges (due to the high allowable stresses), but also higher than originally intended because of the mistake made during the design (when the self-weight was underestimated). After trying to resign Cooper never again visited the bridge site, instead from then on he supervised the construction from his office in New York.

In late summer of 1907 – three years after the start of the construction of the steel superstructure – the cantilever arm reached 223 m out over the river (Fig. 4.8).

The suspended mid-span section (see Fig. 4.2) had temporarily to be fixed to the ends of the two cantilever arms during erection – when the two cantilevers finally were to meet and closed in the middle, then the fixed ends of the suspended span were to be released. In this stage of the construction we have in a way two cantilevers – first the 171.5 m long cantilever arm, and then a 51.5 m long part of the suspended span extending from the nose of the cantilever arm (Fig. 4.9).

Fig. 4.8 The south cantilever arm and the first three panels of the suspended span of the Quebec Bridge extending 223 m out over the St. Lawrence River in late summer of 1907 (see also Fig. 4.9). In comparison to the massive Forth Bridge (see Fig. 3.23) this structure looks for sure anything but "bulky". (Hammond: Engineering Structural Failures – the causes and results of failure in modern structures of various types)

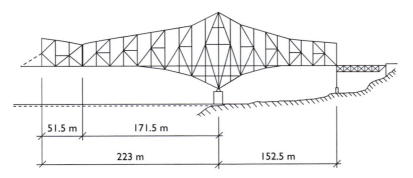

Fig. 4.9 The south half of the Quebec Bridge extending 223 m out over the St. Lawrence River, balanced by the 152.5 m long anchor arm. Three panels are remaining (i.e. 51.5 m) of the suspended span to the mid-point of the bridge (see also Fig. 4.2).

In mid summer of 1907 the first signs of distress in the steel superstructure began to show – a field splice in the ninth panel lower compression chord on the left-hand side of the anchor arm (member A9-L) had become distorted (Fig. 4.10).

While the Forth Bridge had circular tubes the Quebec Bridge had riveted built-up rectangular truss members. The lower compression chord in panel nine of the south anchor arm consisted of four parallel steel plates, $95 \times 1385\,mm^2$ (each consisting of four 23.8 mm thick plates being clamped together by rivets) (Fig. 4.11).

Close to the upper joint of the compression chord in panel nine (at a distance of some 20% of the member length) the web plates were spliced with cover plates (see also Fig. 4.10) (Fig. 4.12).

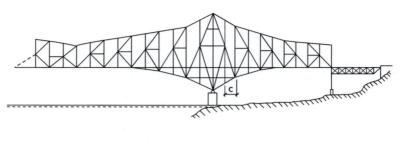

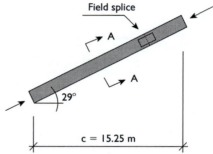

Fig. 4.10 The lower compression chord in the ninth panel of the south anchor arm (member A9-L), where distortion of the field splice was detected in July 1907.

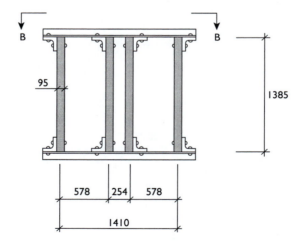

Fig. 4.11 The cross-section of the anchor arm lower compression chord in panel nine (section A-A in Fig. 4.10).

As the lower chords of the anchor arm initially were subjected to tension – when the falsework was removed and the span became simply supported – and the normal force soon changed to compression as the cantilever arm grew outwards, it was decided to temporarily use bolts in order to allow for the structure (the anchor arm) to develop all of its deformations (read: the gradual elongation/compression of the upper and

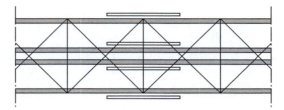

Fig. 4.12 The web splice (spliced in field and not in shop) of the lower chord in panel nine – section B-B in Figure 4.11, i.e. the top view (however, rotated 90°). Also shown are the stabilizing members in the transverse direction – crossing diagonals and horizontals of small L-profiles.

lower chords) before riveting the plates tight together. The anchor arm was quite simply judged to be allowed to adjust itself to the changing load, otherwise secondary constraints could have been developed. However, by taking this measure the buckling strength of the lower compression chords became affected negatively. Besides having observed local distortion of the plates in the field splice, it was discovered in the beginning of August 1907 that chord A9-L was slightly bent out-of-plane as well. The representatives of the Phoenix Bridge Company claimed that the deformations were the result of an incident which had taken place in 1905, when this particular chord had been dropped and bent in the storage yard. Bridge inspector McLure – Cooper's assistant at the bridge site – was, however, convinced of that the deformations had taken place *after* that the chord had been installed, and he was supported by Cooper who suggested that it was most probably a hit damage during construction. In the discussions it was more a matter *when* the chord had been bent rather than if this in any way could affect the safety of the bridge. The concentration was also focused on the construction of the cantilever arm, and not to the more or less already completed anchor arm, so the matter was perhaps not given the highest priority.

At the end of August – when the cantilever had been extended further out over the river, increasing the stresses even more in the truss members – it was found that the out-of-plane bending of chord A9-L had increased from some 20 mm to alarmingly 57 mm. Even though this deformation – representing an out-of-plane deflection of L/305 (the length L of the member being 17.44 m) – is hard to detect by the naked eye, it was a clear signal that something was quite wrong. And visible signals were also given, as rivets for the transverse bracing members (the diagonal and horizontal L-profiles – see Fig. 4.12) began shearing off – a clear evidence of the excessive out-of-plane bending taking place. Realizing the danger McLure immediately travelled to New York on the 27 August to consult with Cooper (telephoning was judged to be too risky – there was always the possibility that the telephone operator would listen in and reveal the situation to the newspapers). Cooper, realizing the danger of the situation, wanting the construction to come to an immediate halt, telegraphed to the Phoenix Bridge Company: "*Add no more load till after due consideration of facts*". However, the contractor (i.e. the Phoenix Bridge Company) was under great pressure not to delay the construction, so the work continued by unfortunately moving an erection crane one step further out on the cantilever. If they had listened to the message from Cooper,

Fig. 4.13 The collapsed 152.5 m long south anchor arm of the Quebec Bridge. The photo is taken from the main pier and facing south towards the river bank. The cantilever arm and the part of the suspended span that was completed lay under water on the river bed. (www.civeng.carleton.ca/ECL/reports/ECL270/Introduction.html)

and acted thereupon – e.g. by immediately removing as much load as possible from the structure, strengthening the bent chord and possibly also supporting the cantilever arm in some way – then perhaps the bridge could have been saved, instead it was now beyond the point of no return.

In late afternoon of 29 August 1907 – just some 15 minutes before the working day was over – the entire bridge collapsed (Fig. 4.13), starting with the buckling failure of the A9-L lower chord of the south anchor arm.

Of the 86 workmen that were on the bridge at the time of the collapse 75 were killed. Had the structure failed just some fifteen minutes later, or above all, had the construction work been stopped two days earlier as was demanded by Cooper, then these lives could possibly have been saved.

Following the collapse, an inquiry concluded that the failure of the bridge was due to the combined effect from the following factors:

- An underestimation of the self-weight loading,
- Temporary splicing of the chords using bolts,
- The allowable stresses being too high, resulting in weak cross-sections,
- The stabilizing bars (the crossing diagonals and horizontals) were not strong enough to resist out-of-plane buckling of the compression chords.

Quite simply, the findings of the inquiry could be summarized as follows: The load-carrying capacity had been overestimated, and the loading underestimated!

There was also much criticism given of how the entire project had been managed:

> "It was clear that on that day the greatest bridge in the world was being built without there being a single man within reach who by experience, knowledge, and ability was competent to deal with the crisis."

If Cooper would have been present at the bridge site during construction – especially during the last months when the trouble began to show – then most certainly the problems would have been dealt with in a different and more competent manner, however, his mere involvement in the project had created "*a false feeling of security*", as the inquiry stated.

We will in the following study the buckling resistance of the lower compression chord and see why and how it failed. If we look at the cross-section (see Fig. 4.11) we have – with respect to the stiff direction of the individual web plates – a possible buckling mode in the in-plane direction (i.e. in the vertical direction, parallel to the plane of the web plates). We could calculate the critical buckling load (having a buckling length, L_{cr}, equal to the length of the member, L) according to (Eq. 4.1):

$$P_{cr}^{in\text{-}plane} = \frac{\pi^2 \cdot EI}{L_{cr}^2} = \frac{\pi^2 \cdot 2.1 \cdot 10^8 \cdot 4 \cdot \left(\dfrac{95 \cdot 1385^3}{12}\right) \cdot 10^{-12}}{17.44^2} = 573.3 \text{ MN} \quad (4.1)$$

This is the critical buckling load according the classical Euler theory (well-known to all engineers in the beginning of the 1900's), assuming an ideal and perfect strut having no imperfections whatsoever. In design this value is used to find a certain slenderness parameter, which then governs the actual load-carrying capacity of the member, where initial imperfections and residual stresses are taken into account.

If we now calculate the critical buckling load in the other direction, i.e. in the transverse out-of-plane direction relative to the web plates, then the value will be (assuming full composite action) (Eq. 4.2):

$$P_{cr}^{out\text{-}of\text{-}plane} = \frac{\pi^2 \cdot 2.1 \cdot 10^8 \cdot 13.5 \cdot 10^{-2}}{17.44^2} = 919.9 \text{ MN} \quad (4.2)$$

where the second moment of area is calculated as (neglecting the small L-profiles – "flanges" – connected at the top and bottom to the stiffening trusses – see Fig. 4.11) (Eq. 4.3):

$$I_y = 95 \cdot 1385 \cdot \left(2 \cdot \left(578 + \frac{254}{2}\right)^2 + 2 \cdot \left(\frac{254}{2}\right)^2\right) = 13.5 \cdot 10^{10} \text{ mm}^4 \quad (4.3)$$

Finally we can calculate the critical buckling load for an individual web plate buckling in between the fixed nodes of the stiffening truss, having a buckling length equal to the

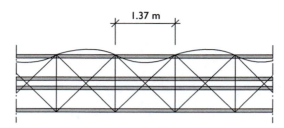

Fig. 4.14 The possible failure mode of an individual web plate buckling independently of the others.

distance between the horizontals (once again neglecting the top and bottom L-profiles for the sake of discussion) (Fig. 4.14 and Eq. 4.4):

$$P_{cr}^{local} = \frac{\pi^2 \cdot 2.1 \cdot 10^8 \cdot 4 \cdot \left(\dfrac{1385 \cdot 95^3}{12}\right) \cdot 10^{-12}}{1.37^2} = 437.1 \text{ MN} \qquad (4.4)$$

If we now compare these three values of the critical buckling load, we see that global buckling in the transverse out-of-plane direction is the *least* critical (assuming full composite action), instead these calculations show that the individual plates buckling locally in between the fixed nodes of the stiffening truss is the most likely to occur (however, neglecting the influence of the "flanges"). This might perhaps explain why the recorded displacement in the global out-of-plane direction was not regarded as critical – but still it was in this direction that the lower chord eventually A9-L did fail.

If we somewhat underestimated the critical buckling load for the individual web plates (by not considering the longitudinal L-profiles at top and bottom) – a true value should probably be in the neighbourhood of that of the in-plane buckling value – we *overestimated* the same (i.e. the critical buckling load) for global out-of-plane buckling (even though not considering the longitudinal L-profiles – however, the influence on the second moment of area is here even less than for the local buckling case). If we are to assume composite action – i.e. the four web plates acting as a unit when being deformed in the out-of-plane global direction – then we need a stiffening truss that is sufficiently strong and robust, and that does not deform easily in its in-plane direction (Fig. 4.15).

However, there is no such thing as full and complete composite action. As lateral out-of-plane deformations are starting to take place then normal forces in the stiffening truss members are also being initiated (as the stiffening truss becomes activated), resulting in axial deformations (elongations and shortenings of the stiffening truss members) which results in reduced composite action between the web plates. We could summarize: The more heavy members used for the transverse bracing, the less axial deformations of its individual members, resulting in higher composite action – and the opposite – the more smaller members used for the transverse bracing, the more axial deformations of its individual members, resulting in less composite action. However, in the beginning of the 1900's this was not fully understood. The transverse bracing member dimensions were based on results from tests on small-scale specimens carried out some 20 years earlier by Bouscaren. In these tests the bracing members chosen excluded

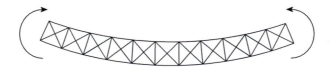

Fig. 4.15 The in-plane *bending stiffness* of the two plane trusses (located on top and bottom of the four web plates – see Fig. 4.12 – as two separate truss units) is the parameter that governs the capacity of the chord to carry a normal force with respect to out-of-plane buckling – not only by ensuring composite action between the web plates, but also by resisting the tendency for out-of-plane deformations (following the relationship: The higher the bending stiffness the lower the curvature, for one and the same bending moment). The capacity of the chord in the vertical direction is not helped at all by this latticing (as the stiffness of the latticing is negligible in this direction).

the possibility of local buckling failure and also ensured adequate composite action for the global buckling strength. When the built-up compression chords of the Quebec Bridge were designed the web plates were scaled up in order to meet the required dimensions (for the needed load-carrying capacity), however, the transverse bracing members were not scaled up to the same extent – it was as if these members were of minor importance, the mere presence of the same was enough to ensure full stability. Unfortunately no full-scale buckling tests were performed to verify this assumption.

Modern research has shown that there is a direct relationship between the size of the stiffening truss and the critical buckling load of a built-up compression chord (Eq. 4.5):

$$P'_{cr} = \cfrac{1}{\cfrac{1}{P_{cr}} + \cfrac{d^3}{E \cdot A_d \cdot a \cdot b^2}} \tag{4.5}$$

where:
P'_{cr} reduced critical buckling load (out-of-plane direction) taking *incomplete* composite action into account
P_{cr} critical buckling load assuming full and complete composite action
d length of the diagonal; $d = (a^2 + b^2)^{0.5}$
E modulus of elasticity
A_d area of the diagonals – in one panel – in the stiffening trusses (i.e. at the top and bottom of the web plates)
a distance between the horizontals
b width of the chord

If we perform a parametric study of this expression above, we find that there is *no reduction* of the critical buckling load, i.e. we have full composite action, when the second part of the denominator approaches zero, i.e. when the bending stiffness of the stiffening truss is very high (Eq. 4.6):

$$\cfrac{1}{\cfrac{1}{P_{cr}} + 0} = P_{cr} \tag{4.6}$$

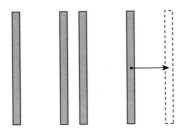

Fig. 4.16 The global buckling capacity of four web plates having no interaction whatsoever is almost equal to zero.

And the opposite, i.e. when we have the second part of the denominator approaching infinity, when the bending stiffness of the stiffening truss is very low, then the critical buckling load is *reduced to zero* (Eq. 4.7):

$$\frac{1}{\dfrac{1}{P_{cr}} + \infty} = 0 \qquad\qquad (4.7)$$

This latter result is equivalent to the case when the web plates are free to buckle individually of each other in the global out-of-plane direction – as four separate units – i.e. when there is no stiffening truss at all (Fig. 4.16).

We see from equation 4.5 that the key parameter that governs the bending stiffness of the stiffening truss – when we keep the material and the geometry parameters constant – is the area of the diagonals, just as we earlier have discussed. We can also identify the axial stiffness of the diagonal, being $E \cdot A_d / d$.

When the axial stiffness of the diagonals decreases (read: when the area of the same decreases) then there is less composite action between the web plates joined together, and the bending stiffness not only of the stiffening truss, but consequently also that of the chord decreases – and as the deformations increase (due to a reduced stiffness) the web plates behave less like a compact unit. We see from expression 4.1 that the key parameter governing the critical buckling load is the bending stiffness EI, and as the bending stiffness decreases so does the critical buckling load.

If we now apply this equation (4.5) to the conditions of the lower chord of the Quebec Bridge we find that there is a markedly reduction in the critical buckling load in the global out-of-plane direction when we take the stiffness of the bracing into account (or rather: the lack of the same) (Eq. 4.8):

$$P'_{cr} = \frac{1}{\dfrac{1}{919.9} + \dfrac{1.97^3}{2.1 \cdot 10^5 \cdot 6452 \cdot 10^{-6} \cdot 1.37 \cdot 1.41^2}} = 316.6 \text{ MN} \qquad (4.8)$$

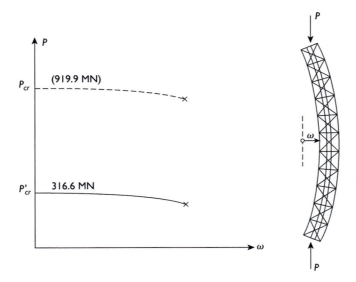

Fig. 4.17 Load/displacement-diagram showing the critical buckling loads in the global out-of-plane direction for full and complete composite action (P_{cr}) and for the case when the flexibility of the stiffening truss is taken into consideration (P'_{cr}).

$A_d = 4 \cdot 1613 = 6452 \, \text{mm}^2$ the total area of four L-flanges in one panel (the diagonals) – two at the top and two at the bottom – where each diagonal is L $100 \times 75 \times 9.5 \, \text{mm}^3$ having a cross-section of $1613 \, \text{mm}^2$

$d = (1.37^2 + 1.41^2)^{0.5} = 1.97 \, \text{m}$

In a graph we could also visually see that the critical buckling load in the global out-of-plane direction has been reduced to almost one third of the value assumed by Cooper (Fig. 4.17).

What has not been shown – neither in figure 4.15 nor in figure 4.17 – are two cover plates used as extra reinforcement in the end panels (instead of having the diagonals), however, this extra stiffness was counteracted by the fact that temporary bolting at the joints was used during construction (instead of riveting).

Both these values of the critical buckling load, P_{cr} and P'_{cr}, belong to theory alone, and are imaginary values assuming a perfect and ideal chord without any imperfections whatsoever. The member – according to the theory – is perfectly straight and does not bend until the critical buckling load is reached, and then it collapses. For any given value of the normal force below the critical buckling load it is possible to apply an additional horizontal (transverse) load – the axially loaded member will deflect, but will, as soon as the applied horizontal load is removed, deflect back. The closer the normal force is to the critical buckling load the smaller this capacity for an additional

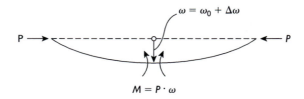

Fig. 4.18 Additional bending in an axially loaded strut due to initial out-of-straightness.

horizontal load becomes, and when the critical buckling load is reached then the member just barely balances the normal force – any disturbance in the horizontal direction will initiate collapse (read: $\omega \to \infty$).

In real design the magnitude of the critical buckling load in relation to the maximum possible normal force capacity given no instability risk (the yield strength times the cross-sectional area, $f_y \cdot A$) will govern the load-carrying capacity of the member. The *higher* the critical buckling loads the less probability that buckling should occur, hence *more* of the chord's full capacity can be utilized. And the opposite, the *lower* the critical buckling loads the higher probability that buckling should occur, hence *less* of the chord's full capacity can be utilized.

In contrast to the theory – where the initial transverse deflection ω is assumed to be zero, and remains so until P_{cr} has been reached – real structural members (such as struts subjected to compression) are always *imperfect* (with respect to shape and material), which makes the load/displacement-relationship differ from that of the ideal theory. For example does the inevitable out-of-straightness make the behaviour become softer and the ultimate load to become much lower. As soon as a load is applied upon an imperfect strut – having a certain initial out-of-straightness ω_0 – there will be an increase of the deflection due to so-called second order effects, so that a bending moment will be produced (Fig. 4.18).

The out-of-plane displacement ω will continue to grow as the applied load is increased, thus producing additional bending stresses which are added to the stresses produced by the axial normal force (Fig. 4.19).

The effect is the worst on the concave side (see Figs. 4.18 and 4.19) where the combined effect of the compressive stresses is the highest. In comparison to the behaviour of an ideal strut real struts will experience a softer load/displacement-relationship because of this effect, and the ultimate load will become much lower than the elastic critical buckling load. There will be an earlier initiation of yielding which lowers the stiffness (and the load-carrying capacity) of the strut – the outer parts will become "soft" due to the yielding, because the stress/strain-relationship of steel is horizontal after yielding has occurred. The ultimate load is consequently becoming lower as the initial out-of-plane displacements increases (as the initiation of yielding starts earlier).

In addition to this combined effect of stresses coming from the normal force and the secondary bending, there are also residual stresses within the material to consider. Residual stresses are in-built self-balancing stresses because of the manufacturing process – a hot-rolled steel plate will experience a difference in contraction as the material cools, the inner parts cool the slowest and are resisted to contract to the full by the

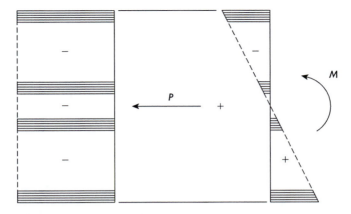

Fig. 4.19 The stress distributions for to the applied axial normal force and the additional bending in the out-plane-direction.

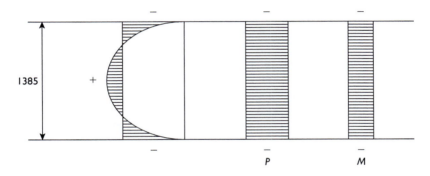

Fig. 4.20 Principal distributions of the residual stresses together with the stress distributions produced by the normal force and the out-of-plane bending for the uttermost web plate on the concave side.

already hardened and stiff outer parts (i.e. the edges). This difference in cooling rate will produce residual (i.e. remaining) tensile stresses in the inner parts and counterbalancing compressive stresses (close to the magnitude of the yield stress of the material) at the edges (Fig. 4.20).

In consequence to the discussion above regarding the superimposition of the stresses the residual stresses will also affect the load-carrying capacity. The residual compressive stresses at the edges will result in early first yielding in these particular parts (if the plate is compact enough that is, otherwise the initiation of local buckling will take place). However, as the buckling direction that we study is in the out-of-plane direction (relative to the face of the web plates) this influence from the residual stresses is not so drastic. The residual compressive stresses at the edges of the web plate is affecting the load-carrying capacity and stability more negatively in the vertical in-plane-direction – the stiffness of the built-up section is governed by the area alone in the out-of-plane direction and not by the second moment of area as it is in the in-plane direction (see

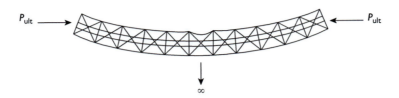

Fig. 4.21 Local buckling of the web plate on the concave side triggered most probably the collapse.

expressions 4.1 and 4.3). In design of today the position of the residual compressive stresses (relative to the buckling direction) together with the magnitude of the same (i.e. the steel quality) are taken into consideration, as well as initial imperfections such as out-of-straightness (up to certain given tolerance levels).

For the case of the built-up sections of the Quebec Bridge yielding of the web plate at the concave side (for out-of-plane bending) was probably not the triggering factor for the collapse of the chord, instead local (Euler) buckling in between the horizontals of the transverse stiffening truss was more probably the initiating cause – starting with the web plate on the concave side having the highest combined effect of normal force stresses and secondary bending stresses (besides the presence of the residual stresses) (Fig. 4.21).

As local buckling of the web plate on the concave side took place then there was nothing to stop the collapse of the chord (the remaining three web plates in the damaged panel were unable to carry the load as the eccentricity of the normal force increased drastically). The whole loading procedure was just like a gigantic full-scale load test until failure; a step by step increase in load and a simultaneous monitoring of the deformations, just as it is done in a laboratory – the only difference being that here it was performed in field conditions, in a real load-carrying structure. Starting with an already damaged (read: initially bent) chord there was a gradual increase in loading – and out-of-plane deflections – until the ultimate load-carrying capacity was reached (Fig. 4.22).

Knowing what we know today of the stability of built-up chords – and also knowing what we know of the outcome – the maximum design load-carrying capacity should only have been a fraction of the ultimate load, however, at that time when the Quebec Bridge was built Cooper was convinced that the capacity was sufficient (relying upon full composite action). The stiffening truss was expected of having adequate strength and stiffness, but instead it was its weakness and lack of rigidity that made the chord (and the bridge) to collapse. As the out-of-plane deflection of the chord increased the strain on the stiffening truss members increased. Each diagonal and horizontal had only a single rivet at each end, making the shearing stresses (as well as the bearing pressure) becoming high, as the loading and the deflections of the chord increased. Prior to the collapse there were also reports of rivets shearing off, and that decreased the load-carrying capacity (and the stiffness) of the stiffening truss even more. The final failure of the chord could also have been preceded by the rupture/buckling of some of the stiffening truss members.

Due to the lack of rigidity in the field splice – by the temporary bolting – the stiffness of the chord could also very well have been weakened so that the collapse was triggered

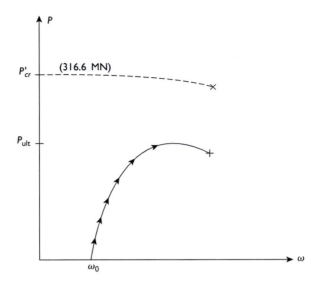

Fig. 4.22 The true load/deformation-relationship (in principal), well below that of the theoretical response assuming an ideal and perfect chord. The actual loading procedure was just like following a gigantic load test until failure.

Fig. 4.23 The possible forming of a hinge in the compression chord caused by the temporary bolting of the field splice.

by mechanism behaviour, i.e. by the forming of a hinge in the field splice affecting the stability not only in the vertical in-plane direction but also in the transverse out-of-plane direction (Fig. 4.23).

In contrast to Dee, Ashtabula and the Tay Bridge – which all failed first after some years in service – the Quebec Bridge did not even reach its full completion. However, the difference being that the Quebec Bridge was subjected to heavy strain already during construction because of the assembly method (not a problem for the Forth Bridge though). The lower chord A9-L – already bent due to mishandling at the storage yard and having a reduced load-carrying capacity because of this – became more and more deformed as the loading increased, not only due to the initial out-of-straightness, but also due to the lack of strength and rigidity of the stiffening truss. This sideways bending was observed, but they did not really know how to handle with it, and they did not fully understand the severity of the situation – Cooper's assistant at the bridge site, McLure, tried his best to convince the contractor of the danger of the situation, but it was too late as the ultimate load had already been reached. When the chord failed due to sideways (out-of-plane) buckling there was nothing to stop the collapse of the entire structure – as it was statically determinate the loss of one single member transformed it within seconds from a load-carrying structure to a mechanism, i.e. a body set in motion.

The four separate web plates of the chord would readily have been able too carry the loading, not only during construction, but also in the completed bridge (taking full service load), *given* that they interacted to the full through composite action that is. And the opposite; they would not have been able to carry any load at all (more or less) without transverse stiffening (i.e. acting as four free and separate plates). The actual condition of the chord was somewhere in between these two extremes, probably closer to the latter. Cooper had overestimated the critical buckling load of the lower compression chord member, and therefore also overestimated the design load-carrying capacity of the same. The attention lay on the primary members and not on the secondary – Cooper had assumed that the stiffening trusses were of "minor importance" (relying entirely on the small-scale tests performed by Bouscaren) and did not perform any full-scale tests of his own to verify this assumption. Perhaps his instincts told him that there was not any need for a strong and robust stiffening truss as the compression chords were not intended to bend (straight as they were assumed to be). Perhaps he saw that wind loading on the face of the chords was the only loading producing bending, and that is more or less negligible with respect to stresses produced in the stiffening truss members. One can more or less suspect that the stiffening truss members were chosen at random. But then, no regulations regarding the design of stiffening trusses existed at that time. The whole project was also characterized by quick decisions and keeping the costs down to a minimum (to save money and to beat the Forth Bridge with respect to the amount of material used). For sure Cooper was perhaps a little bit too confident in his own abilities (having had success in all the projects during his career and being praised as one of the leading engineers of that time). Phoenix Bridge Company was also renowned for producing quality bridges (at a low price), and their reputation together with Cooper's gave all those involved in the project a false sense of security – there was no need of having somebody outside of the project double-checking the design (at one stage during the construction process though this was proposed, but Cooper strongly opposed to the idea). And that Cooper supervised the project from his office in New York gave the signal to the people at site that everything was safe. To sum it all up; everybody involved in the project was convinced that everything was okay, and there was nothing at that time that proved that something was wrong, except the lack of verification that the stiffening trusses were adequate that is.

Following the recommendations of the inquiry, extensive tests were carried out on riveted built-up columns after the collapse of the Quebec Bridge. It was found that when a stiffening truss system similar to that of the Quebec Bridge was used the columns failed at a much lower load than was predicted. By increasing the member dimensions of the stiffening truss, and by having two rivets at each end of the members (thus reducing the shear on the rivet, and also giving a more fixed-end condition for the members), the load-carrying capacity of the columns increased markedly. These were findings that is in accordance with the knowledge that we have today of the influence which a flexible stiffening truss system has on the stability, but at that time, in the beginning of the 1900's, this was new knowledge – it is a typical example of learning through failure. A deficiency in the general knowledge became painfully identified. It was also written in the inquiry's report (supporting Cooper to a certain extent) that the design of the stiffening truss for riveted built-up compression chords:

> "...depends on the judgment of the engineer guided solely by experience. He finds little or nothing in scientific text books or periodicals to assist his judgment."

But as the case is for any new structure to be built – containing structural elements where the knowledge concerning their load-carrying ability can be questioned or is somewhat diffuse – there is *always* a need for verifying tests. In a comparison though between the Forth Bridge – where the testing was extensive – and the Quebec Bridge – where the testing was scarce (to say the least) – one must not forget that in the former case the extra precautions taken were a direct result of the Tay Bridge collapse. Had Cooper experienced a similar failure in his previous career he had most certainly also required some extra tests on the Quebec Bridge.

Even if it was an important finding of the immediate testing that followed the Quebec Bridge collapse – i.e. to choose strong and rigid members for the stiffening trusses in built-up chords – the influence from initial out-of-straightness and residual stresses still remained to be studied, and so it was done in the following decades to come.

In the aftermath of the Quebec Bridge collapse two famous bridges over the East River in New York – namely the Manhattan Bridge and the Queensboro Bridge, both being under construction – were examined a little closer. The Manhattan Bridge was a suspension bridge with a main span of 448 metres, and had the Phoenix Bridge Company as contractor. The task of checking the specifications was given to none other than Gustav Lindenthal, the bridge engineer that drew up the plans for just a suspension bridge over the St. Lawrence River – a proposal that was rejected by the Quebec Bridge Company. But Lindenthal was not free from being scrutinized himself. He was the designer of the Queensboro Bridge – a twin-cantilever truss bridge in five spans (the longest being 360 metres) – and it was found that the live-load capacity had to be adjusted down, because the dead load (i.e. the self-weight) had been miscalculated (have we not heard that before!). Both the Manhattan Bridge and the Queensboro Bridge were successfully opened in 1909.

Regarding the failure to bridge the St. Lawrence River at Quebec, it was already in 1908 decided to make a second attempt. This time there would be a group of competent bridge engineers working side by side, instead of relying just upon one single individual. Among the group were names like Ralph Modjeski (chief designer of the Manhattan Bridge) and Maurice Fitzmaurice (which had participated in the construction of the Forth Bridge) – Cooper was not contracted, as he had retired (at the age of 68) after having testified before the Royal Commission of Inquiry.

The new bridge – which began to be constructed in 1909 – was more or less similar in shape to the old bridge. The upper and lower chords were straight (instead of being curved) and the arm lengths were slightly altered, but the main span was still the same (i.e. 549 m). This time the design and work were done in a most meticulous way:

- Higher design live loads were adopted,
- Accurate and precise calculation of the self-weight,
- Allowable stresses at a reasonable level,
- The design calculations were checked and double-checked,
- Strong and robust stiffening trusses for the compression chords,

Fig. 4.24 The suspended span lifted free from the barges and hanging in between the two cantilever arms. (Hammond: Engineering Structural Failures – the causes and results of failure in modern structures of various types)

- Extensive testing of both material and members,
- Extra care in handling,
- Avoiding temporary joining using bolts,
- But above all, no saving of material if not necessary.

The result of the precautions taken in the design, resulted in a 150% (!) increase in cross-sectional area for the compression chords in relation the old bridge (even though a steel of a higher quality – i.e. with higher allowable stresses – was used for the new bridge). But still – even though this second attempt to raise the bridge was paid the utmost focus and concentration on maximum safety – more troubles lay unfortunately ahead.

It had been decided to lift up the central suspended span from a barge (or six of them, to be exact), instead of – as in the first bridge – having one half of the suspended span reaching out from the end of the cantilever arm. In this way the maximum free cantilever length became reduced, thus lowering the maximum compression (and tension) chord forces during erection. The suspended span, 195 metres long – slightly shorter than in the original bridge – having a weight of more than 5000 tons, was to be lifted up in position – 46 meters above the water level – at least, that was the intention. In September 1916 the construction of the suspended span was completed, and it was then shipped to the bridge site and floated in underneath the two cantilever arms. After the barges had been safely anchored, the lifting operation could commence. First the four corner ends of the suspended span had to be connected to the lifting hangers, and then hydraulic jacks could begin to lift up the span in steps of 60 centimetres at a time (Figs. 4.24 and 4.25).

When the suspended span had been lifted 9 metres up above the water level there was a crash due to the sudden collapse of one of the corner bearings. Immediately the eccentric loading became too much for the remaining bearings to carry (more or less

Fig. 4.25 A close-up of the hanger arm and one of the corner bearings. (Hammond: Engineering Structural Failures – the causes and results of failure in modern structures of various types)

Fig. 4.26 The second failure in September 1916 in the construction of the Quebec Bridge (quite a remarkable picture taken by a very alert photographer). (www.civeng.carleton.ca/ Exhibits/Quebec_Bridge/intro.html)

twice the loading for the intact diagonal corner bearings), so the entire span crashed down into the river, twisted and deformed (Fig. 4.26).

This time 13 workers were killed (in addition to the 75 already lost in the first collapse in 1907). There were many theories put forward for the reason of the failure,

Fig. 4.27 The Quebec Bridge over the St. Lawrence River in Canada, the longest cantilever bridge in the world.

but it was finally attributed to inadequate strength of the connection detail due to a faulty design.

Due to the sudden unloading of the cantilever arms (having been deflecting down almost 20 centimetres due to the weight of the suspended span), they started to oscillate violently, but were left undamaged. Of the suspended span, however, nothing remained more than wreckage lying on the bottom of the river.

The decision was taken to make a second attempt – persistent as they were to close the gap between the two cantilever arms – and a year later (almost to the day), in September 1917, a duplicate span was successfully being lifted up in position, having the connection detail between the hangers and the corner bearings completely redesigned. The project, which had started almost 20 years earlier, was finally completed and the bridge could at last be opened for traffic. In October 1917 the first train crossed the bridge, and one could imagine the tension among the spectators (and for the engine driver of course) hoping that nothing would go wrong.

The bridge is still in service today, and remains actually, with its main span of 549 metres, the longest cantilever bridge in the world (Fig. 4.27).

Cooper died on 24 August 1919 (at the age of 80), and lived to hear about the opening of the ill-fated bridge, however, not participating in the actual completion himself, of what originally was intended to become the jewel in his crown.

And finally, going back to the comparison between the Forth and the Quebec Bridge – the former being a huge success and the latter a failure (before completion that is) – one must remember that we learn from both successes and failures, and even though the Forth Bridge was regarded as a success, it was also not spared from casualties – 57 workers died in the construction of that bridge.

Chapter 5

Hasselt Bridge

On a cold morning in March 1938, the Hasselt Road Bridge in Belgium suddenly and unexpectedly collapsed and fell into the Albert Canal, less than two years after its completion (Fig. 5.1).

The bridge was of the Vierendeel type (invented by the Belgian engineer Arthur Vierendeel), having a span of 74.5 metres (Fig. 5.2).

A Vierendeel bridge is somewhat similar in shape to that of a tied arch/bowstring arch bridge – which is having vertical hangers between the arch and the lower horizontal member (the tie) – but instead the Vierendeel bridge should be categorized as a

Fig. 5.1 The collapse in March 1938 of the Hasselt Road Bridge over the Albert Canal in Belgium. (Stålbygge. Utdrag ur Stålbyggnad, programskrift 11)

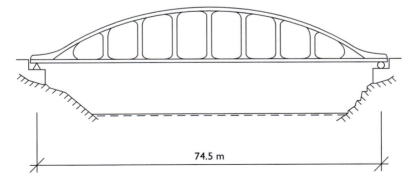

74.5 m

Fig. 5.2 The Vierendeel bridge over the Albert Canal near the city of Hasselt, close to the border of Germany, some 70 kilometres east of Brussels, the capital city of Belgium.

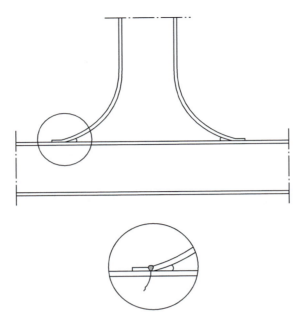

Fig. 5.3 The initiating point where the crack started that ruptured the lower tension chord.

framework, acting more or less like a truss without diagonals, having stiff connections between the members in order to carry load as a rigid frame.

When investigated it was found that collapse of the bridge was initiated by the fracture of the lower chord, starting with a crack originating from a transverse butt weld in the transition between the concave gusset (the haunch) and the horizontal lower chord (Fig. 5.3).

The Hasselt Bridge was an all-welded structure having box-girder sections for all the members, and the fracture had made the lower member crack in a brittle manner. A number of other small cracks in other parts of the bridge were also found. The primary cracking – making the tension chord to be severed – was most certainly starting from a small fatigue crack, originating from the butt weld in the joint – the failure came when the crack had reached its critical length with respect to brittle fracture.

In the winter of 1940, two more welded Vierendeel bridges in Belgium – at Kaulille and Herentals Oolen – also experienced extensive cracking, however, not collapsing as the Hasselt Bridge did. Two years prior to the collapse of the Hasselt Bridge, a welded I-girder railway viaduct over the Hardenberg Strasse in Berlin, Germany, had to be replaced as longitudinal cracks had been discovered in the longitudinal weld between the flange and the web. Another viaduct in Germany experiencing problems was the continuous road bridge at Rüdersdorf outside Berlin, which in 1938 had its main girders rupturing by brittle fracture. Both these German bridges had their failures occurring during winter time, in cold weather, just as the three Belgian bridges.

There are a number of influencing factors that makes steel – at the presence of a notch (read: stress raiser – such as a weld, a fatigue crack, or a sharp change in

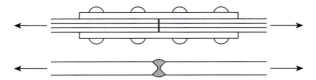

Fig. 5.4 A riveted butt joint and the equivalent welded solution.

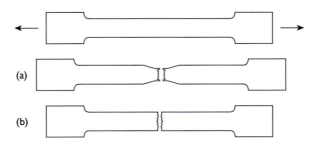

(a)

(b)

Fig. 5.5 A tension test, where the fracture response is (a) ductile, and (b) brittle.

geometry) – respond in brittle rather than a ductile manner (i.e., losing its capacity of plastic strains), and these early welded bridges in Belgium and Germany, experiencing problems with brittle fracture cracking, more or less sums up all of these factors. These factors will shortly be discussed, but first a brief comment upon why welding – in the 1930's – became the dominating joining technique in bridge construction, replacing riveting.

The introduction of steel in bridge engineering (Forth Bridge, Quebec Bridge) made it possible to develop another joining technique than riveting, and arc welding was used for the first time in the late 1920's to strengthen and repair bridges. In the mid 1930's all-welded bridges began to be constructed, and in comparison to an equivalent riveted structural solution the advantages are apparent (Fig. 5.4).

In the welded solution no extra cover plates in the splice are needed, thus saving material. Instead of having several thinner plates, thicker plates were also used, which excluded the need of having individual plates clamped together. In all, great savings were made (less material and quicker assembly). But the switch in the 1930's from riveting to welding for the construction of steel bridges initiated the problem with brittle fracture, something that earlier – in riveted steel bridges – had not been a problem.

The ability to plastify – rather than to fracture in a brittle manner (i.e. experiencing unstable crack growth) – is the single most important factor that makes steel such a safe and unique construction material (besides its high strength of course). When a steel structure is overloaded, and localized yielding is taking place, the stresses normally are redistributed to other more stiff areas (read: to areas which still respond elastically), thus unloading the overstressed region. However, under certain conditions – e.g. at the presence of sharp notches – this characteristic can be markedly reduced.

If we consider a standardized tension test, there is a significant difference in result whether the steel samples respond in a ductile or brittle manner (Fig. 5.5).

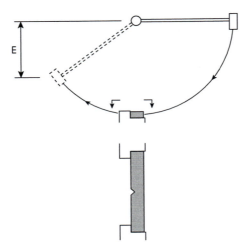

Fig. 5.6 The Charpy V-notch impact testing procedure.

The specimen that responds in a ductile manner undergoes large plastic (irreversible) strain before fracturing (hence, becoming visibly elongated). The initial fracture is a "shear failure" within the material, i.e. the slippage of dislocations (incomplete atom planes inside the crystal) located at an angle of 45° to the loading direction (dislocations located at an angle of 0° or 90° are unable to slip), while the final fracture of the narrowing net section is brittle. The specimen that responds in a brittle manner fails due to the cleavage (breaking) of the atomic bonds, perpendicular to the loading direction, without prior warning (read: without any yielding of the material).

The factors that govern whether the fracture should be ductile or brittle, are temperature, loading rate, plate thickness, steel quality, ageing, welding and fatigue. These parameters will be discussed one by one in the following.

I TEMPERATURE

Lowered temperature is the most apparent common denominator for the bridges in Belgium and Germany that failed in the late 1930's.

Under certain conditions – when the temperature is normal and the loading rate is low – steel is capable of elongating some 20% or more, but when subjected to impact loading, at the presence of notches, and at cold temperatures, it can fracture in a brittle manner (i.e. without elongating). When the temperature is low, and the steel contracts, there is quite simply less ability for the dislocations to slip as the material has been compacted.

In the Charpy impact test (named after its creator, the French scientist Georges Charpy) one can study the temperature influence on the fracture behaviour of steel. Small machined V-notched specimens are struck by a pendulum hammer (released from a fixed height) (Fig. 5.6).

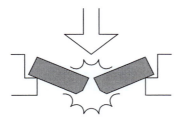

Fig. 5.7 The pendulum strikes the steel specimen, which breaks in two.

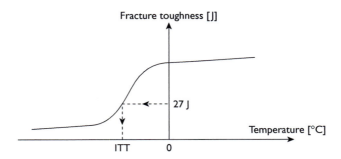

Fig. 5.8 A typical result of the toughness/temperature-relationship found at Charpy testing. The impact transition temperature (ITT) is marked – defined as the temperature when the fracture energy is 27 J (Nm).

The loss of swinging height E for the pendulum is equivalent to the fracture energy required to break the specimen – the specimen is positioned so that it will break in three-point bending, having the V-notch on the tension side (of course) (Fig. 5.7).

As the testing temperature is lowered the fracture toughness decreases (read: the pendulum height increases due to the gradual loss of the specimens to retard the swing – the plastic deformation of the specimens decrease, and thus the resistance), and at a certain range there is more or less a distinct drop of the fracture energy (Fig. 5.8).

In many codes it was suggested that the ITT – i.e. where the fracture behaviour, according to the results from Charpy impact testing, switches from ductile to brittle – should be below the lowest in-service temperature, but this requirement is way to crude (but in some sense quite logical). There are many other influencing factors which have to be taken into account (e.g. what is the true plate thickness in the existing structure? – more on the influence of the plate thickness follows – and is it probable that notches of such a size are present in a bridge, subjected to such high impact loading weight and rate?). When tested, steel samples from old bridges have had an ITT of +5°C or above (in an environment where the lowest winter temperature could be expected to be −30°C or below), but still having 80–100 years of in-service loading without any fracturing tendency. The answer lies in the correlation between plate thickness and the probability to respond in a ductile or brittle manner. For thin plates the ITT should be adjusted downwards, giving the lowest allowable temperature in service, while for thick plates it should be adjusted upwards. See more about the influence of the plate thickness in the corresponding section below.

2   LOADING RATE

At Charpy impact testing the pendulum hammer weight and its velocity are parameters that are kept constant (as the temperature is the variable to be studied), but would the impact energy be increased then the fracture energy would have been decreased (for a given temperature). The ITT would in such a case be raised, i.e. the transition between ductile and brittle fracture behaviour would occur at a higher temperature, i.e. the probability of a brittle fracture increases with an increase in loading rate. Both the yield point and the tensile strength increases with an increase in loading rate (the former more than the latter), having the effect that the steel will respond elastically at higher load levels, thus more easily initiate cleavage fracture rather than shear fracture. There is less available time for the dislocations to slip as the loading rate increases. In a tension test it would show as a reduced ability to elongate plastically (as the yield point and the tensile strength are increased).

Rapid changes in temperature could also have the same negative effect on the capacity to respond in a ductile manner, as an increased loading rate has. At the presence of a sharp notch (such as an existing crack) a fast temperature drop could very well initiate cleavage fracture (without there being any additional external load); as a matter of fact that was the case in the Rüdersdorf Bridge. Just before the brittle fracture of its bridge girders the temperature fell from 0° to −12°C within a very short period of time. As the material is fast contracting there is not only a reduction in the capacity to respond in a ductile manner, but also a high probability that high tensile stresses are (rapidly) produced because of an uneven temperature change (the cooling rate could differ between different parts and also within the material – the surface cools faster than the interior). The quicker this change in temperature the faster the material has to adjust. A slow and even temperature contraction does not produce any stresses.

An example of quite the opposite, but still very much related to the discussion above, is the fact that an ordinary drinking glass, containing a small crack, will crack into pieces when put in hot boiling water. Here it is not the rapid heating that is the main reason for the breakage; the glass would have broken anyway because of its inability to plastify – a brittle fracture is initiated at the crack tip because of the expansion of the glass material – a typical example of a fracture because of temperature deformations (in combination with inner constraints which produces tension stresses at the crack tip).

3   PLATE THICKNESS

The thicker the plate the higher the probability is of a brittle behaviour. As the thickness of a steel plate increases, there is a higher concentration of carbon and slag impurities in the inner central parts, and this is because of the uneven cooling rate after the rolling procedure. Because of the slow cooling in these inner parts there is also the forming of larger grains (having had a longer time to develop). Larger grains and the higher concentration of carbon and slag inclusions (oxides, sulphides) are the reason why these inner parts show a less ductility when tested. There is also a "size effect" which has to be taken into account – the probability of a defect (a large slag inclusion, a void, or a crack) is very much higher in a thick plate than in a thin. And even if there is the same probability of a defect being present a built-up plate (consisting of several layers of thinner plates), than in a single thick one (see e.g. Fig. 5.4), the defect in the former case is only affecting the individual plate – it would fracture alone without the entire failure of the built-up plate, while in a single thick plate a fracture would mean the entire collapse of the plate. This difference in fracture behaviour has often been observed in fatigue tests of riveted built-up members.

In thick plates – subjected to tensile loading – there is also a high probability of tri-axial tensile stresses being produced around a notch. As plastic flow is being initiated (due to the stress concentration effect) – and plastic stress redistribution could reduce the peak stress at the notch – the surrounding material (which still is responding elastically) will resist the contraction at the notch, not only in the in-plane transverse direction (relative to the loading direction), which is the case for thinner plates, but also in the thickness direction of the plate. These additional tensile stresses in the thickness direction (also those transverse to the loading direction) resist the forming of plastic flow at the notch, lifting the stresses up *above* the yield level. Cleavage fracture will thus be more easily initiated in thick plates where bi-axial stress states easily could turn into tri-axial stress states. In the German bridges that failed, thicknesses of up to 60 mm were used.

4 STEEL QUALITY

When the content of carbon in steel increases the tensile strength also increases, however, normally at the cost of a reduced ductility (i.e. the capacity of undergoing plastic strain, prior to fracture, is lowered). The content of sulphur and phosphorus are also affecting the ductility negatively. As the ductility is decreased – because of a high content of carbon, sulphur and phosphorus – the ITT is raised (which of course is unfavourable).

New high-strength steel was developed in Germany at the middle of the 1930's, having a tensile strength of 520 MPa. The increase in strength helped to save material in the new bridges that were built (e.g. Hardenberg Strasse and Rüdersdorf), however, because of the high carbon content (in combination with the high allowable stresses) the material was not suited to be used in welded bridges. The content of carbon in this high-strength steel was later reduced to 0.20% and sulphur and phosphorus to 0.10%. In modern weldable steels the content of manganese is high – at the same time as the carbon content is kept low – so that the ductility is not affected even though the strength is high.

5 AGEING

Under certain conditions steel could "age", meaning the deterioration with time with respect to its ductility. Especially Thomas steel (basic Bessemer) is susceptible to this phenomenon. In the basic Bessemer process the lining of the converter is basic (rather than acid) using Dolomite rock – which captures the acid phosphorus oxides – making it possible to use high-phosphorus pig (raw) iron. The process was inexpensive so the steel price became low, making it popular. However, as the Bessemer method used air (instead of pure oxygen) in the decarburization process (when the high carbon content coming from the pig iron is reduced through oxidation) the content of nitrogen also became high. Free nitrogen atoms can then, with time, diffuse out of the (Thomas) steel and "lock" to the ends of the dislocations (where there is free space in the iron lattice), thus making the steel harder and more brittle (the steel becomes less capable of shear failure within the material). The effect of ageing is especially marked after a cold-deformation process (read: when the steel has been subjected to plastic deformations) because of the high number of new dislocations being produced. The heat shock treatment during welding can also be a factor that speeds up the diffusion rate. All three bridges that experienced brittle fracture in Belgium were made of Thomas steel.

6 WELDING

In contrast to riveted joints – which are capable of small adjustments when subjected to excessive loading (due to the slipping of the joined plates) – welded joints are more stiff and rigid, which reduce its ability to redistribute stresses. Thus there is a higher probability for localized overstressed regions in a weld, behaving as a sharp notch as it already is. Besides the stress concentration effects, there is the presence of in-built residual tensile stresses (due to the restricted contraction during cooling – the higher the steel quality the higher the residual stress level) and weld defects (incomplete penetration and fusion, the inclusion of gas pores, and small cracks), which are all parameters that make the welded region more susceptible for fatigue and brittle fracture (increased grain size and the concentration of carbon in the heat-affected zone are also parameters that contribute). In the welded bridges that experienced brittle fracture problems during the late 1930's, the use of high-hydrogen electrodes (which increased the embrittlement of the weld), together with groove welds, poor design and execution, and wrong welding sequence (introducing large deformations and additional constraints), all contributed to the initiation of the brittle fractures. As was mentioned in the section about steel quality, the content of carbon was eventually reduced in the German high-strength steel (making it more suitable for welding), but there was also the introduction of low-hydrogen electrodes and the prohibition of using thicker plates ($t > 20$ mm) in welded joints without preheating.

7 FATIGUE

In order to avoid fatigue cracking, the notches there are (read: the stress raising points) should be kept as smooth as possible. Any sharp change in geometry, however small, will create localized stresses that are very much higher than the nominal stress. In a weld a fatigue crack is easily initiated from any defect within the material or from the roughness of the surface. The cracking at the initial stages are small localized shear failures inside the material (irreversible plastic strains at the notch) when the joint is subjected to fluctuating stresses back and forth. In order to open up and fracture the material the fluctuating stresses need to be tensile, however, there is also the probability of crack initiation in a compressed zone because of the presence of residual (tensile) stresses. The varying magnitude of the nominal compressive stress will make the residual tensile stress fluctuate, but as soon as a crack has been initiated, the continued propagation of the same will be stopped because of the release of the residual stresses.

Even though fatigue cracking is something that has to be avoided – as long as possible anyway – it is, in relation to brittle (cleavage) fracture, a safe failure mode – the crack just grow slowly for each new loading cycle. However, at a certain critical length, the stress concentration effect at the crack tip becomes so high that brittle fracture takes place – this particular length is shorter in cold climate, because of the brittleness of the material in cold condition, and that is why the net-section (brittle) failure normally takes place during the winter.

In old and worn structures, where corrosion is present, fatigue cracks could also be initiated in otherwise smooth regions because of the general loss of area (raises the nominal stress level) and – which is worse – the creation of small holes (pits) on the surface which act as local stress raisers.

Going back to the failure of the Hasselt Bridge in Belgium, this Vierendeel bridge concept was in a way quite provident, as the transition between the verticals and upper and lower chords was rounded, providing a smooth flow of forces (see Figs. 5.1 and 5.2). However, because of the framework action, large localized stresses were being formed due to the combined effect of bending and normal forces.

The design stresses for the Hasselt Bridge (as well as for the other welded bridges that failed during the late 1930's) – taking the lowered fatigue strength of welded joints into account – were kept too high (assuming the same fatigue resistance as riveted connections) because of the lack of fatigue test results. The welding technique, as well as its strength, was new and unproven, therefore the lack of fracture toughness was something that was not enough considered. Once again – in the bridge engineering history – it was an example of learning by failing! And with respect to fatigue (leading up to brittle fracture), the knowledge of why cracks do occur, and where they are being initiated – i.e. where to look for the same – was something that was not common knowledge among the bridge engineers, as riveted bridges had not been experiencing these kinds of problems to such an extent before.

These brittle fracture failures in Belgium and Germany in 1936–1940 more or less sum up all of the influencing factors that govern whether the fracture response should be ductile or brittle:

–   Cold weather,
–   Thick plates,
–   High-strength and brittle steel (because of high carbon content),
–   Groove welds,
–   Poorly executed welding, and bad welding technique in general,
–   Bad detailing (with respect to position of the welds),
–   High residual stress level, coming not only from the welding itself, but also from improper welding sequence leading up to large welding deformations and high additional constraints,
–   Hydrogen embrittlement due to the use of high-hydrogen electrodes,
–   Dynamically loaded structures (high strain rates and fluctuating stresses initiating fatigue cracks),
–   The use of Thomas steel, which is especially prone to ageing,

–   High allowable stresses (by using steel of a high quality or by neglecting the lowered fatigue strength of welded joints).

And as these factors were present in the bridges (not necessarily all in one and the same bridge), the elastic strain energy at the notch (normally a fatigue crack) was lifted to such a level that cleavage fracture of the material was initiated.

The welding technique, the proper making of fracture tough steel and the understanding and learning about fatigue in welded joints, were developed in the years to come after these failures.

Chapter 6

Sandö Bridge

When the decision was taken in 1937 to bridge the river Ångermanälven at Sandö – and to replace the existing ferry – the choice fell upon a concrete arch bridge, which was to become the longest in the world. The river is located some 450 kilometres to the north of Stockholm, the capital city of Sweden (Fig. 6.1).

Fig. 6.1 The Sandö Bridge is located at the outlet of the Ångermanälven into the Baltic Sea.

Fig. 6.2 The timber arch falsework being shipped into position. (www.famgus.se/Gudmund/ Gudm-Sandobron.htm, with kind permission of Leif Gustafsson)

Construction of the falsework began in April 1938, and this 247.5 metre long timber arch was to be a self-supporting formwork during casting of the concrete. In May 1939 the timber arch was completed and shipped into position (Fig. 6.2).

By choosing a self-supporting timber arch as falsework during the construction of the concrete arch bridge, it becomes quite clear that in order to build a bridge you normally need to start by first building a bridge (i.e. a temporary one)!

The shape of the arch was chosen so that the pressure line for evenly distributed load would coincide with its centre line. A horizontal tie bar ensured the stability of the timber arch during the transport.

The $11.3 \times 4.0\,m^2$ cross-section of this temporary timber bridge construction consisted of two compact timber flanges, separated by a large number of internal lattice walls (Figs. 6.3 and 6.5).

The timber lattice walls was connected and fitted to the flanges through nailing, and the flanges – also nailed together – consisted of more than 220 individual pieces of timber, each having the dimension of $50 \times 200\,mm^2$ (Fig. 6.4).

Every six metres the box was stabilized in the transverse direction through cross-bracing walls (Fig. 6.5).

The lattice walls consisted of over-lapping timber elements, $50 \times 150\,mm^2$, and they were to function as diagonal elements in a regular truss system, ensuring equilibrium during bending between the normal forces in each joint (Fig. 6.6).

However, as the bending moments were negligible – because of the dominating compression – the diagonal member forces were small.

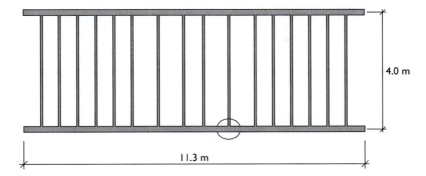

Fig. 6.3 The cross-section of the timber arch falsework for the Sandö Bridge.

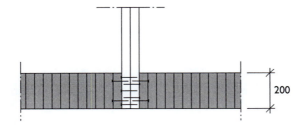

Fig. 6.4 The built-up bottom flange and internal timber wall connection.

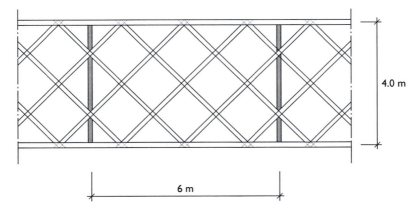

Fig. 6.5 Elevation of the timber arch.

As mentioned earlier, the Sandö Bridge was to become the longest concrete arch bridge in the world, eclipsing the Plougastel Bridge in France, which was a three-span concrete arch bridge (each span 187 metres) built in 1930 and constructed by the famous engineer Eugène Freyssinet (inventor of prestressed concrete). Freyssinet also used a self-supporting timber arch falsework during the construction of his bridge.

Immediately after the timber arch had been put into position the concrete casting began, and the work proceeded during the entire summer of 1939 without any problems. In

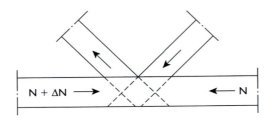

Fig. 6.6 The normal force equilibrium in each joint.

order to minimize the bending deformations of the falsework the casting was done at several locations on the arch, at the north and south side simultaneously. On 31 August, only a short section at the crown of the arch remained to be cast. The workmen had for several days experienced some uncomfortable vibrations coming from the bridge, but in the late afternoon this particular day a noise from within the structure suddenly grew into a large roar, as the bridge began to break down and collapse. As the bridge fell into the river, a thunder just like an earthquake was heard for miles around. A huge surge flooded the shores, and all the nearby houses had their window panes crushed. Within just a matter of a few seconds this enormous structure, under the self-weight of the timber arch and the additional weight of the concrete, collapsed leaving nothing but emptiness behind. Of the 30–40 workmen that still were on the bridge, 18 died. An account from one of the survivors:

> "The only thing I remember was that all support was lost underneath my feet and I fell into an empty void. The huge timber arch fell in front of me as I tumbled down into the water. I was fully conscious and felt the blow as I hit the water, thinking that I must not get any water in my lungs. Suddenly I felt a blow to my chest and saw a concrete cart pass in front of me down in the water."

Next day, on 1 September 1939, when the newspapers wrote about this catastrophe, the front pages were dominated by the news about the outbreak of the Second World War – Hitler had declared war on Poland. The news about the collapse of the Sandö Bridge did not reach the big headlines and it was soon forgotten by the general public.

An expert commission was summoned to investigate the causes behind the collapse, and they declared that some of the transverse cross-framing members, and possibly also the diagonals in the lattice truss system, had inadequate cross-sectional dimensions. It was also suggested that the compressive stress in the timber arch contributed to the collapse, however, at the time of the collapse, under the weight of timber and concrete, the stresses were in the neighbourhood of 6–7 MPa, so that was a highly improbable contributing cause. It was finally stated that the exact cause could not be established. Professor C. Forssell suggested that the collapse was due to lateral (sideways) buckling of the nailed timber flanges of the arch, but this explanation was rejected.

It was not until some 20 years later, in 1961, that the exact cause was put forward, and it was shown that Professor C. Forssell was correct in his assumptions. Professor Hjalmar Granholm, at Chalmers University of Technology in Göteborg, had devoted some time and effort in bringing light to the cause, and he could prove that the bending

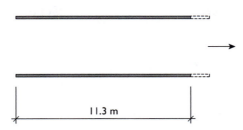

Fig. 6.7 Buckling direction, i.e. in the stiff direction of the flanges.

stiffness of the built-up flanges were too weak to resist lateral buckling. If we – for the sake of discussion – start by calculating the bending stiffness of the two flanges (as a coupled pair) in order to find the critical buckling load, assuming *full interaction* between the joined timber parts, we find (Eqs. 6.1 and 6.2, Fig. 6.7):

$$I_o = 2 \cdot \frac{0.2 \cdot 11.3^3}{12} = 48.1 \, \text{m}^4 \tag{6.1}$$

$$\Rightarrow \quad EI_o = 10,000 \cdot 48.1 = 481,000 \, \text{MNm}^2 \tag{6.2}$$

where the modulus of elasticity, E, for timber has been taken as 10,000 MPa.

The critical buckling load, q_{cr}, we find by assuming the flanges as two plates – buckling in their stiff direction – having a buckling length equal to $L_b/2$ (fixed end conditions), where L_b is the arch length (Eq. 6.3):

$$H = \frac{q \cdot L^2}{8 \cdot f}$$

$$\Rightarrow \quad q_{cr} = \frac{8 \cdot f}{L^2} \cdot \frac{\pi^2 \cdot EI}{(0.5 \cdot L_b)^2} \tag{6.3}$$

$$H_{cr} = \frac{\pi^2 \cdot EI}{L_{cr}^2}$$

where L is the (horizontal) length of the span, and f the rise of the arch.

If we use the actual values for L and f (see Fig. 6.8), we finally have the critical buckling load, where we have assumed full interaction between the individual timber elements in the flanges (Eq. 6.4, Fig. 6.8):

$$q_{cr} = \frac{8 \cdot 36.5}{247.5^2} \cdot \frac{\pi^2 \cdot 481,000 \cdot 10^3}{(0.5 \cdot 262)^2} = 1318.7 \, \text{kN/m} \tag{6.4}$$

We continue now by comparing the critical buckling load value to the actual loading just before the collapse. The self-weight of the arch was 1000 tons (metric) and the weight of the poured concrete was 3000 tons (Eq. 6.5):

$$q = \frac{(1000 + 3000) \cdot 9.81}{247.5} = 158.5 \, \text{kN/m} \ll q_{cr} \tag{6.5}$$

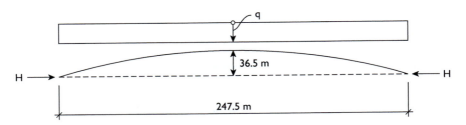

Fig. 6.8 The span length and rise of the timber arch.

However, as there is nothing like full composite action in a built-up section – especially not for a nailed timber flange – the *true* second moment of area must be used. What we have is something like a glulam timber beam, however, without the glue. Granholm performed a large number of tests in order to establish the shear deformation characteristics of the nailed joints in a built-up timber beams similar to the Sandö Bridge timber arch flanges, and he found that the second moment of area is only some 13% of that of the unreduced value (where full composite action had been assumed) for immediate and *short-term* response. The true critical buckling load value therefore becomes (Eq. 6.6):

$$q_{cr}^{short} = 0.13 \cdot 1318.7 = 171.4 \, \text{kN/m} > q \tag{6.6}$$

Even though the reduction was large, the actual load is still below the critical buckling value, however, the margin is far too small to be accepted in a design situation (as the critical buckling load is a theoretical value assuming perfect and ideal conditions).

But there are more reductions to consider. Granholm found that the *long-term* response is such that the deformations increase, i.e. that the bending stiffness is lowered. Creep (i.e. the increase in deformations at constant loading) in the timber material as well as in the nailed joints (the shearing deformations of the nails produce high localized contact pressure zones) lowers both the modulus of elasticity and the second moment of area. The long-term values should therefore be 7,000 MPa (instead of 10,000) and $0.10 \times I_0$ (instead of $0.13 \times I_0$), giving the critical buckling load value to (Eq. 6.7):

$$q_{cr}^{long} = 0.10 \cdot \frac{7,000}{10,000} \cdot 1318.7 = 92.3 \, \text{kN/m} < q \tag{6.7}$$

This last result indicates quite strongly that lateral buckling was the cause of the failure of the timber arch. We have, however, neglected the stiffening effect from the already hardened concrete, but also not considered the reducing effect on stiffness and strength coming from the climate (rain and wet concrete increased the moisture content in the timber) – local bending effects of the top flange due to the weight of the poured concrete has also not been considered. But all these additional factors taken together (strengthening *and* weakening) do not alter the result that global buckling in the lateral direction was the cause of the failure. The inevitable presence of an initial out-of-straightness must also have produced additional bending stresses that lowered the buckling strength further.

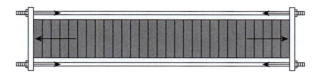

Fig. 6.9 Compacting through clamping would have increased the shear resistance.

Fig. 6.10 The timber falsework used for the second (and this time successful) attempt to bridge the river Ångermanälven at Sandö.

Fig. 6.11 The Sandö Bridge over the river Ångermanälven in Sweden.

Perhaps some sort of applied clamping pressure would have saved the bridge (by increasing the composite action of the built-up timber flanges), however, hard to perform (Fig. 6.9).

The expert commission – that suggested that possibly also the compressive stresses were a contributing factor behind the collapse – was in a sense correct, however, not having lateral buckling in mind.

The bridge was finally rebuilt in 1943, however, now using a timber falsework supported all along its length (Fig. 6.10).

The strength of the timber material was tested meticulously and even the number of knots (!) was counted.

The bridge is still carrying traffic today, and remained until 1964 the longest concrete arch bridge in the world with its main span of 264 metres (Fig. 6.11).

Since the opening in 1943 the crown of the arch has settled more than 500 millimetres due to shrinkage and creep deformations – concrete as well as timber are for sure living materials!

Chapter 7

Tacoma Narrows Bridge

The collapse of the Tacoma Narrows Bridge is perhaps the best recorded and documented bridge failure in the bridge engineering history – the spectacular and prolonged failure process was captured on extensive live footage, giving a unique document for the investigation committee as well as for the engineering society at large. The footage has since then been used in civil engineering classes all around the world for educational purposes, and it is a very instructive video showing the consequences of neglecting dynamic forces in the construction of suspension bridges.

The bridge was taken into service in July 1940, and was located over the Tacoma Narrows in the Puget Sound some 60 kilometres south of Seattle in the State of Washington (Figs. 7.1 and 7.2).

It was constructed as a 1660 metre long suspension bridge with a main span of 853 metres (see Fig. 7.2), the third longest suspension bridge in the world (and USA)

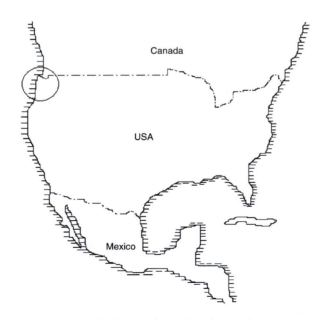

Fig. 7.1 The Tacoma Narrows Bridge was located in the northwest corner of the USA, in the State of Washington.

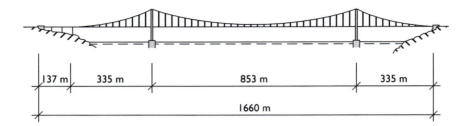

Fig. 7.2 Elevation of the Tacoma Narrows Bridge.

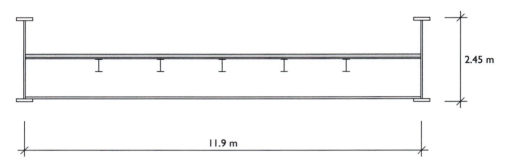

Fig. 7.3 The stiffening girder (i.e. the bridge deck) of the Tacoma Narrows Bridge, which provided room for two traffic lanes and foot pavements.

at that time. The bridge deck was 11.9 metre wide, supported by two 2.45 metre deep plate girders (I-girders) (Fig. 7.3).

Ever since the introduction of iron suspension bridges in the beginning of the 1800's the stability in wind has been a problem for the designers. The deck (i.e. the stiffening girder) was not given sufficient stiffness as the traffic load of that time was comparatively low (to modern standards). When Thomas Telford's Menai Strait Suspension Bridge in Wales – the longest bridge in the world at that time (177 m) – was damaged in a storm in 1826 (the same year it was built) it was concluded that this is as far as one can build. The maximum length for suspension bridges – with respect to stability and safety – was reached (at least, that was the conclusion). But even smaller bridges suffered from wind damages. In 1836 the Royal Suspension Chain Pier in Brighton ("Brighton Pier") collapsed. Its four main spans of each 78 metres were completely destroyed in a heavy wind. The same year the Menai Strait Bridge suffered damages during a storm, despite being reinforced in 1826 after its first damages. Three years later, in 1839, the Menai Bridge once again suffered some severe damages because of excessive wind.

The flexibility of the bridge decks (read: their lack of stiffness) caused not only problems with vibration and swaying during wind loading, but also, for example, when marching troops were passing. Through the combined effect of heavy wind and the steps interlocking with the eigenfrequency of the bridge, a large troop of marching soldiers in 1850 set the suspension bridge over the river Maine at Angers in France in violent vibrations. The bridge collapsed and 226 soldiers lost their lives. This tragedy had a major impact on suspension bridge design, and for example on the Albert Bridge, built in 1873 over the river Thames in London, a posted sign was saying: "*All troops must break step when marching over this bridge*" (the sign is still there to read!).

The main watershed came, however, in 1854, after the Wheeling Suspension Bridge over the Ohio River, some 80 kilometres southwest of Pittsburgh, in the State of West Virginia (some 100 kilometres south of Ashtabula), collapsed during a storm. Charles Ellet, the designer of the Wheeling Bridge, had in 1849 been awarded the contract to build the bridge, having presented a more slender and therefore more economical solution than the proposal presented by John A. Roebling. Ellet strongly advocated flexibility in the stiffening girder and that the large self-weight – because of the record-breaking length (308 metres, being the longest bridge in the world) – would ensure a stable construction without interfering with the forces of nature (i.e. wind). Roebling had suggested a bridge with shorter spans and a more robust stiffening girder. The Wheeling Bridge was after the collapse reconstructed, adding more stiffness to the girders, and was later also strengthened with inclined stays by the Roebling Company. John A. Roebling became after this the leading bridge designer for suspension bridges for the years to come, and he designed stiff and robust suspension bridges, crowning has career in 1883 with the 486 metre long Brooklyn Bridge in New York (the longest in the world, and the first suspension bridge where steel wires were used for the main cable).

However, as the years progressed into the 1900's, it was as if the lesson learnt from the Wheeling Bridge collapse gradually was forgotten. Using robust stiffening girders (normally deep trusses) had more or less ruled out the problems with instability in wind, and therefore there was a tendency to lessen the stiffness of the deck when increasingly longer suspension bridges were being built – wind was "no longer" any problem to consider (as no problems had occurred), and weight had to be saved in order to maximize the load-carrying capacity for traffic (i.e. the live load). The introduction of high-strength steel was also a factor that favoured less material being used (due to higher allowable stresses). In 1931, when George Washington Suspension Bridge was built – being the longest bridge in the world with a main span of 1067 metres – it was originally designed for just one bridge deck, and it was also suggested that deep stiffening trusses should be omitted, however, the future need for a second deck made it necessary to provide the bridge deck with such stiffening elements. Six years later, in 1937, the Golden Gate Bridge was built, breaking the record of George Washington Bridge – the main span being 1280 metres. The bridge deck of Golden Gate Bridge was also stiffened by deep trusses.

During this period the deflection theory began to influence the design of suspension bridges, as it provided a tool for the distribution of live load in between the stiffening girder and the main cable in a more economical way than before – with respect to static load-carrying capacity alone it was not necessary to carry so much load by the stiffening girder as earlier had been done. As a suspension bridge is a statically indeterminate structure any solution in which the load is distributed in between the stiffening girder and the main cable is a valid solution. In fact, the solution where the bridge deck is having absolutely no stiffness at all is also a possible solution; however, this leads to an extremely flexible and swaying bridge, suitable only for shorter pedestrian bridges for occasional passages (one person at the time). In order to distribute concentrated loads, and to resist excessive lateral deflections during wind loading, the bridge deck (i.e. the stiffening girder) needs to be not only deep but also wide, i.e. having an adequate bending stiffness in both the vertical and horizontal (lateral) direction.

By the introduction of the deflection theory, and the long period of not experiencing any major problems with wind, the designers were led to believe that the stiffening girder could be chosen only according to the requirement that concentrated live load should be distributed in the longitudinal direction according to some minimum deflection requirements – and by using this theory it was possible to construct more economical suspension bridges than before. The deflection theory had great potential, but it was also tempting to use this theory more than was safe with respect to the stability in wind. With respect to the Tacoma Bridge there was also another circumstance that contributed to the swaying tendencies. The width of the bridge was only 11.9 metres (see Fig. 7.3) because of the fact that only two traffic lanes (one in either direction) was judged necessary. The bridge authorities recommended that the width of suspension bridges should be at least a minimum of 1/30 of the span length in order to ensure a certain minimum lateral stiffness. The Tacoma Bridge had a width-to-span ratio of only 1/72 (11.9 m/853 m), but had been checked against static wind loading and found to be sufficient in both stiffness and strength (with respect to maximum allowable lateral deflection and stresses). The width-to-span ratio of the George Washington Bridge was almost exactly 1/30 (36 m/1067 m) and the Golden Gate Bridge 1/47 (27 m/1280 m).

The *depth-to-span* ratio for the stiffening girder in suspension bridges, in the span range of 600–900 metres, was recommended to be above 1/90 (earlier even above 1/40), that is, with a depth of the stiffening girder of at least 1/90 of the span length. The Tacoma Bridge, with its 2.45 metre high girders (see Fig. 7.3), had a depth-to-span ratio of only 1/348. If they would have been to meet this recommendation then the stiffening girder depth should have been at least 9.5 metres (i.e. 853 m/90)! So it is very easy to conclude that the Tacoma Bridge had a very slender and flexible bridge deck, both in the vertical direction (for in-plane bending) and in the horizontal (out-of-plane) lateral direction – it was as if the ideals of Charles Ellet once again was accepted. Leon Moisseiff, the chief designer of the Tacoma Bridge (and the designer of the main cables for the Manhattan Bridge in 1909 – one of two bridges that were especially checked after the Quebec Bridge collapse, see Chapter 4), relied upon the weight of the long span to ensure stability in wind, having the opinion the longer the span the better (i.e. safer), the same opinion that Ellet also held about his Wheeling Bridge.

This flexibility of the Tacoma Bridge – in the in-plane vertical direction – was something that the car drivers experienced as soon as the bridge was taken into service. Even in very small winds the bridge began to oscillate, and this phenomenon had also been noticed already by the construction workers. Special dampers were installed, and the bridge deck was also stabilized by external tie-down cables (as well as inclined hangers at mid-span – more about this later), but to no avail – the bridge still oscillated in the wind. The Golden Gate Bridge had experienced similar oscillations a couple of years earlier, so the undulations were considered harmless and the traffic continued without any major concern. In fact, the oscillating bridge became something of a tourist attraction, and was given the nickname "Galloping Gertie" (after the "galloping" girders). Instead of frightening the drivers it attracted people from far away who wanted to take a slow and safe "roller-coaster" drive across the bridge, experiencing vertical undulations of up to 1.5 metres! However, in the beginning of November in

1940, only some four months after the opening of the bridge, the conditions suddenly became serious as the situation suddenly went out of control. The following description of the event is taken from the book "Bridges and their builders", by David B. Steinman & Sara Ruth Watson, except for the photos (see Picture and photo references at the back):

"From about 7 A.M. on the morning of November 7, 1940, the bridge had been persistently undulating for some three hours. A wind of thirty-five to forty-two miles per hour (16–19 m/s) was blowing, and the waters of Puget Sound were whipped into whitecaps. The segments of the span were heaving periodically up and down as much as three feet (0.9 m), with a frequency of about thirty-six cycles per minute (0.6 Hz). Alarmed at the persistent character of the wave motion in the span, the highway authorities stopped the traffic over the structure. At 10 A.M., the last truck was passing over the span, when something seemed to snap, and, suddenly, the character of the motion changed. The rhythmic rise and fall changed to a two-wave twisting motion, with the two sides out of phase. The main span was oscillating in two segments, with nodes at mid-span. As two diagonally opposite quarter-points were going up, the other two diagonally opposite quarter-points were going down. The frequency was fourteen cycles per minute (0.23 Hz) and, soon after, changed to twelve cycles per minute (0.2 Hz). With each successive cycle, the motion was becoming greater, until it had increased from three feet (0.9 m) to twenty-eight feet (8.5 m)! At one moment, one edge of the roadway was twenty-eight feet higher than the other; the next moment it was twenty-eight feet lower than the other edge. The roadway was tilted forty-five degrees from the horizontal one way, and then forty-five degrees the other way. Lamp standards in one half of the span made an angle of ninety degrees with lamp standards in the other half. Fortunately, some amateur photographers were at the scene with motion picture cameras, and they have supplied us with a unique and unprecedented record of the action of the span in its dance of death." (Fig. 7.4).

"The motion pictures of the twisting span are unforgettable, and the distortions they depict, in character and magnitude, are almost unbelievable. The span twists in gigantic waves, and it is difficult to realize that the girders were made, not of rubber, but of structural steel having a modulus of elasticity of 29,000,000 pounds per square inch (200,000 MPa)." (Fig. 7.5).

"For a half hour and more, the steelwork and concrete slab took this terrific punishment. Something was bound to give way. At 10:30 came the first break: one floor panel at mid-span broke out and dropped into the water 208 feet (63 m) below. The twisting, writhing motion continued. Spectators on the shores were herded to a safer distance away from the span. At 11 A.M. the real breaking up of the span occurred; 600 feet (183 m) of the main span near the west quarter-point tore away from the suspenders, the girders ripping away from the floor like a zipper; part of the falling bridge floor turned upside down before the entire falling mass hit the water, sending up spray to a great height." (Fig. 7.6).

"With a 600-foot section of the bridge gone, the engineers of the bridge structure expected the motion to subside. But the heaving and twisting of the rest of the bridge continued, with the side spans now participating in the motions. Finally, at 11:10 A.M., nearly all the rest of the main span tore loose and came crashing down. The 1,100-foot (335 m) side spans, now deprived of the counterbalancing weight of the main span, suddenly deflected about 60 feet (18 m), striking the approach parapet; then bounced up with an elastic rebound, only to drop again with a final

Fig. 7.4 The twisting motion of the main span. (University of Washington Libraries, Special Collections, UW 21413)

Fig. 7.5 A still from a classic live film of the twisting span, showing a car left out on the bridge, abandoned by a newspaper reporter. (University of Washington Libraries, Special Collections, UW 21429)

Fig. 7.6 The first part of the collapse, where 183 metres of the main span tore away from the vertical hangers (suspenders) and fell into the waters below. (University of Washington Libraries, Special Collections, UW 21422)

sag of about 30 feet (9 m). This was the final gigantic convulsion in the death struggle of a great bridge." (Fig. 7.7).

Apart from some short ends of the bridge deck only the two suspension cables remained of the main span (Fig. 7.8).

Because of the long duration of the violent oscillations traffic was stopped and the bridge closed well in time of the final collapse, so fortunately there were no casualties.

A committee was immediately summoned to investigate the cause of the collapse. The group consisted of three distinguished engineers: *Othmar Ammann* (chief designer of the George Washington Bridge and also assisting in the design of the Golden Gate Bridge), *Glenn B. Woodruff* (consulting engineer at the George Washington Bridge) and the aerodynamicist *Theodore von Kármán* (a well reputed professor in aeronautics and astronautics). In March 1941, four months after the collapse, they reported:

"The excessive vertical and torsional oscillations were made possible by the extraordinary degree of flexibility of the structure and its relatively small capacity to absorb dynamic forces. It was not realised that the aerodynamic forces which

Fig. 7.7 The deflected side span on the east side, at the time of the collapse of the main span. (University of Washington Libraries, Special Collections, UW 27459z)

Fig. 7.8 The collapsed Tacoma Narrows Bridge – aerial view looking west, a rather sad and dreary sight. (University of Washington Libraries, Special Collections, UW 26818z)

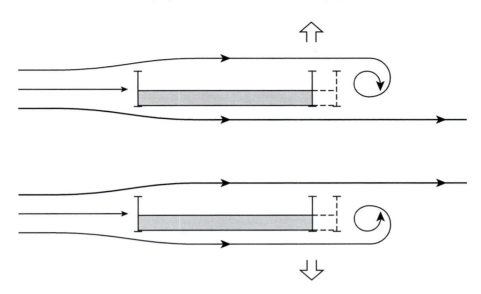

Fig. 7.9 Vortex shedding at a steady air flow.

had proven disastrous in the past to much lighter and shorter flexible suspension bridges would affect a structure of such magnitude as the Tacoma Narrows Bridge, although its flexibility was greatly in excess of that of any other long-span suspension bridge."

The committee concluded that the bridge had been properly designed according to the present codes, and with adequate stiffness and strength for *static* wind pressure – for wind speeds close to three times more than the wind of that fatal morning in November 1940 – however, not taking into consideration the possibility of *aerodynamic instability*. Wind tunnel testing confirmed that the bridge was very susceptible to a phenomenon known as vortex shedding (at least known to aerodynamicists at that time). A cross-section such as the Tacoma Bridge deck – having a small width in relation to its length, and a very flexible stiffening girder, consisting of solid web plates – is especially sensitive to the formation of vortices (whirlpools) in a steady air flow, which appear at regular interval (i.e. at a constant frequency) at the back of the deck (on the leeward side). This phenomenon creates alternating suction in the transverse direction to the wind, perpendicular to the bridge deck, subjecting it to pulling forces up and down (in the vertical direction) at regular frequency, while, at the same time, the bridge is pushed horizontally by the steady wind (Fig. 7.9).

If the shedding frequency of these vortices coincides with the natural frequency of the body, then resonant vibration can be induced, creating excessive oscillations. In the case of the Tacoma Bridge – where the situation was a little bit more complex – these lift and drag forces were initially too small – i.e. to deflect the deck up and down – however, due to the sustained conditions the deflections not only were initiated, but also gradually grew, increasing little by little with time. The investigation committee

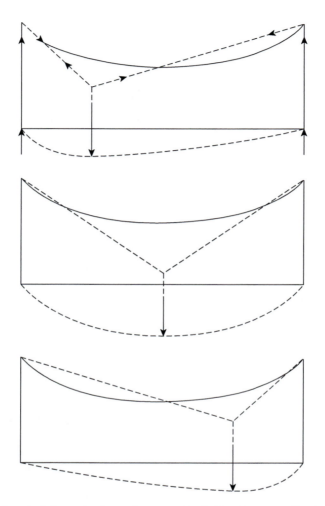

Fig. 7.10 The gradual change in shape of a suspension bridge due to a rolling load.

were able to observe this phenomenon in wind tunnel tests, where a model of the bridge was subjected to a steady air flow, producing both bending and twisting oscillations similar to that of the real bridge.

But before continuing the discussion of different vibration modes (induced by vortex shedding) we will first consider the deformation behaviour under a rolling concentrated load, showing the "deformability" (i.e. softness) of a suspension bridge.

When a suspension bridge is subjected to a concentrated (live) load, the stiffening girder is deflecting while the main cable is adjusting itself to the position of the load (i.e. being straightened), and this adjustment gradually changes as the load is moving over the bridge (Fig. 7.10).

In fact, for short-span suspension bridges (with narrow decks), this rolling "bulge" and the reshaping of the cable form are normally visible for the naked eye when heavy

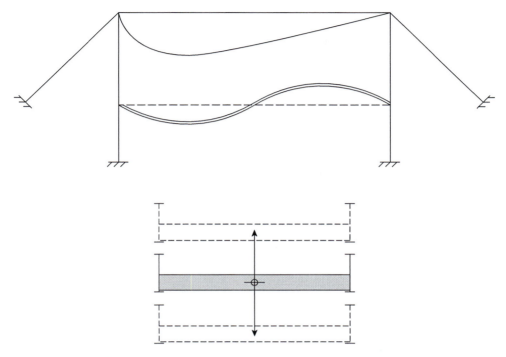

Fig. 7.11 The first bending mode of vibration of a suspension bridge deck (and the corresponding deformation of the main cables).

trucks are passing (at slow to moderate speeds). However, for long-span bridges (having wide decks), this action behaviour is hidden by the fact that the main load is self-weight of the deck, which also governs the shape of the cable – a rolling concentrated load will then just slightly alter the shape of the cable. The heavier the deck, and the less flexible (i.e. more stiff) the stiffening girder is, the less tendency there is for the cable to adjust itself to concentrated loads, but then, this is the way suspension bridges still function. And finally, to conclude; if a suspension bridge is having a light-weight deck, with a flexible stiffening girder, where a large portion of the load (self-weight and live load) is taken by the main cables, then there is high potential for large deflections, just as the case was for the Tacoma Bridge (not only susceptible to wind and traffic induced vibrations, but also soft and deformable for live loads).

And coming back to the wind-induced vibrations we have the primary bending mode, which was initiated in the early phase of that fatal morning on 7 November 1940 (and that also was initiated by traffic and low winds in the months before). In contrast to a simply supported beam – where the first bending mode of vibration is when the entire span is deflecting up and down – the deck of the main span of a suspension bridge is vibrating (oscillating) periodically (i.e. at a fixed frequency) in a sine-wave form, i.e. when half of the span is deflecting up as the other half is deflecting down (Fig. 7.11).

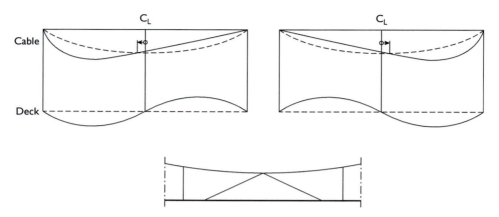

Fig. 7.12 By locking the cable at midspan, e.g. by inclined hangers (centre ties), the vibration of the bridge deck can be limited.

This vibration mode is the form that offers the least resistance for the deformation of the structure (i.e. requiring least energy), as a mode where the entire span is deflecting up and down (as simply supported beams do) would require for the main cable to be strained (i.e. to elongate) – now the cable is only adjusting itself according to the deformation of the deck without any change in length. A good example of the flexibility of the Tacoma Bridge deck and a visualization of the first bending mode of vibration is shown in figure 7.7 of the deflected side span (showing a half-sine wave). The bridge deck is in a local perspective extremely stiff, but, as can be seen, globally very flexible.

One way of limiting these kinds of vibrations is to lock the cable at midspan, as the midpoint of the cable is moving back and forth when the cable is stretched (in relation to the midpoint of the deck that is) (Fig. 7.12).

The main cable of the Tacoma Bridge was in fact locked to the deck in a similar way, so the designers were actually aware of this beneficial effect.

In addition to bending vibrations there is also torsional vibration (i.e. twisting) to consider. The first torsional mode of vibration is when one half of the bridge is rotating one way as the other half rotates the other (Fig. 7.13).

Worth noticing is that the end girders are deforming as two separate girders, vibrating individually in the first bending mode, alternately to each other (the cables are also adjusting each other individually to the deformation). Just as the case was for the first bending mode of vibration for the entire deck there is for torsional vibration also a mode which requires the least energy of the system when deforming.

As the torsional deformation of the deck well has started then there is also a tendency of the wind acting as a driving force, as the surface of the deck is exposed (however, in the opposite direction the wind is stabilizing – pushing down instead of pushing up) (Fig. 7.14).

At the time of the collapse, the oscillations had started as bending vibrations, but *"when something seemed to snap"* (possibly the breakage of the cable locks at midspan) the vertical oscillations changed into twisting.

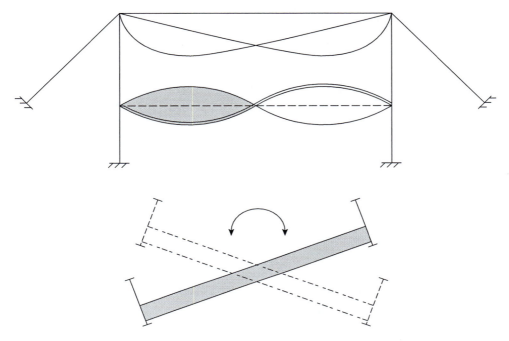

Fig. 7.13 The first torsional mode of vibration of a suspension bridge deck (and the corresponding deformation of the main cables).

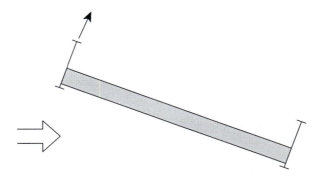

Fig. 7.14 Wind acting as a driving force with respect to torsional oscillation.

A triggering factor with respect to the torsional oscillations – besides vortex shedding – may also have been a phenomenon which is called lateral/torsional buckling. As the wind is pushing the bridge deck horizontally, the two end girders act as two flanges – separated apart by the deck – in a deep beam (which is on the flat). Compression stresses are formed in the end girder at the windward side and tension stresses in the other end girder at the leeward side, and the compression flange could buckle transverse to the wind loading direction (if the wind is strong enough that is) (Fig. 7.15).

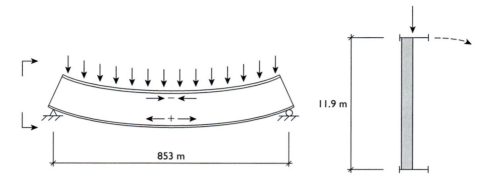

Fig. 7.15 The 853 metre long bridge deck as a simply supported deep beam (should in fact be continuous in three spans), with possible lateral/torsional buckling risk.

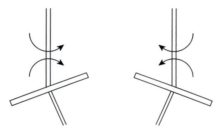

Fig. 7.16 Local bending of the hanger connections as the bridge deck was twisting back and forth made the bridge eventually collapse – a typical case of low-cycle bending fatigue.

The buckling direction of the compression flange would be upwards in the vertical direction (because of the least resistance from the cable), counterbalanced of course by the self-weight of the deck.

Whatever the initiating mechanism of the twisting, there was a constant and persistent driving force coming from the vortex shedding pumping energy into the system, and because of the low structural damping (read: dissipating energy due to the deformations) the torsional oscillations increased step by step with each new loading cycle. Resonance between the vortex shedding frequency and the natural frequency of the bridge deck (for bending vibrations and torsional vibrations) could very well have been the case during some periods, however, as the deformations started the vortex shedding frequency also changed, so it was more like a matter of forced vibration in a system where the motion was self-supporting.

Then, what was it that finally made the bridge deck to break from the vertical hangers and fall into the water below? Of course the bending and twisting deformations of the deck itself were excessive (however, as the flexibility was low no destructive strains were produced), the key failure mechanism was instead the *local* bending of the connection detail between the hangers and the end girders (Fig. 7.16).

As was stated by the investigation committee the bridge had been designed for static wind speeds nearly three times the speed at the day of the collapse, i.e. a wind pressure almost nine times higher (the wind pressure is proportional to the wind speed in square). However, a small secondary and negligible dynamic effect – not known by the bridge engineers – became a destructive force which brought the bridge down. Taking all the primary static loads into account was just not enough, especially as the Tacoma Bridge was given a slenderness and flexibility that exceeded existing suspension bridges with a great margin, making it extremely sensitive to dynamic forces.

In the 1920's and 1930's, when cars and trucks became the major means of mass transport of people and goods, the former requirement of very stiff bridges even for longer spans – that gave structures such as the Forth Bridge and the Quebec Bridge, being railway bridges as they were – was no longer a hindrance to bridge over increasingly longer gaps. The shift from cantilever truss bridges to suspension bridges, due to the change in stiffness requirement, made it possible save material even though longer spans were being built. This temptation, i.e. to save material in the bridge deck, went a step too far in the case of the Tacoma Bridge.

The subsequent wind tunnel testing performed by the investigation committee was not only to confirm and establish the nature of the aerodynamic instability of the bridge (giving answers to questions such as critical wind speed, oscillation amplitude, and structural damping), but also to test new designs. It was found that – not quite surprisingly – that the natural frequency of the bridge deck increased as the bending and torsional stiffness are increased, i.e. that the probability of vortex shedding excitation decreased (higher wind speeds were required). Especially was the stability to dynamic wind forces gained by a wider deck, when the torsional stiffness increases more in relation to the bending stiffness. Having a fixed bending stiffness of the deck the torsional stiffness could be increased by just increasing the width between the end girders. In the case of the Tacoma Bridge it was unfortunate that the circumstances with respect to the traffic volume were such that a wide deck was not required. If future increase in traffic would have been taken into consideration – making room for more than two lanes already from the start – then the torsional stiffness, in particular, would have gained from this measure.

But, above all, it was found that stiffening *trusses*, in contrast to solid plate girders, were especially suited to be used in the design, as they allowed for the wind to flow more or less undisturbed through the structure. The problem with vortex shedding excitation almost vanished when open permeable trusses were used.

After the Tacoma Bridge collapse, when the results of the investigation had been published, the suspension bridges in the USA were examined with respect to their aerodynamic stability. The 701 metre long Bronx-Whitestone Bridge over the East River in New York had the same stiffening girder layout as the Tacoma Bridge – i.e. solid end plate girders – however, a girder depth of 3.35 metres (in comparison to the 2.45 metres in the Tacoma Bridge). The bridge deck was almost twice as wide – 22.6 metres in comparison to 11.9 metres – making it less susceptible to torsional oscillations. However, there had in fact been problems with vertical oscillations, so it was decided to add 4.3 metre deep trusses on top of the girder flanges, thus increasing the construction depth of the end girders to 7.65 metres (3,35 m + 4,3 m). It is worth mentioning that the construction designer was Othmar Ammann, one of committee

Fig. 7.17 The current Tacoma Narrows Bridge. (University of Washington Libraries, Special Collections, UW 7091)

members investigating the Tacoma Bridge collapse – the design of the Tacoma Bridge was thus not unique and extreme (just a bit more slender and flexible than its equal).

In the design of new suspension bridges coming after the Tacoma Bridge the decks were made heavier – relative to the earlier bridges – by choosing a thick layer of concrete, resulting in a higher tension of the main cables (i.e. not entirely counterbalancing the extra weight by an increase of cable dimension). The more the cables are tensioned the less flexible and adjustable to concentrated loads they become, and the stiffer with respect to dynamic wind loading. As a similitude one could say that the cable "strings" were tuned at a higher frequency.

In 1950, ten years after the Tacoma Bridge collapse, the destroyed bridge was replaced by a second bridge in the same location. The design of this new bridge had been thoroughly tested in the wind tunnel laboratory. The solid plate girders of the old bridge were replaced by open permeable trusses (Fig. 7.17).

The bridge deck was 18.2 metre wide, and the stiffening truss depth was 10.0 metres, almost four times as deep as the old plate girders (Fig. 7.18).

The new bridge was exactly as long as the old one, but due to the deeper trusses and the wider deck (and most certainly also a thicker concrete layer) the weight was 50% more. The depth of the trusses also met, in fact, the old recommendation of a minimum depth-to-span ratio of 1/90 ($d \geq 853/90 = 9.5$ m).

For the design of new suspension bridges after 1940 wind tunnel testing became the rule rather than the exception. The solution where bridge decks were given deep stiffening trusses was, however, eventually abandoned. In 1966, when the 988 metre

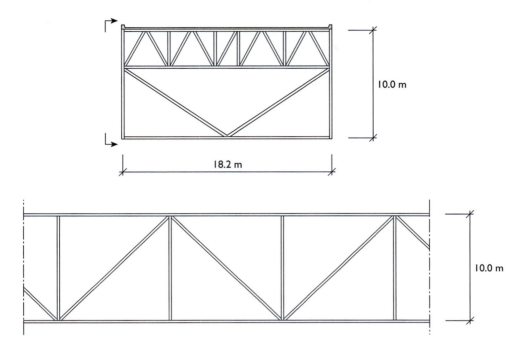

Fig. 7.18 Cross-section of the bridge deck and elevation of the longitudinal stiffening truss (principal layout) of the current Tacoma Narrows Bridge.

long Severn Bridge was built – spanning the river Severn from England to the south of Wales – wind tunnel testing helped the designers to find an optimal streamlined box-girder shape, thus reducing the weight of the deck. Instead of finding the stability by the means of deep stiffening trusses a slender and thin cross-section of the deck was found to be stable in the wind by itself. The saving of material by choosing aerodynamically shaped deck profiles was necessary if longer spans were to be bridged. Deep trusses and heavy decks give stable suspension bridges, however, for longer spans the live load capacity becomes more and more reduced (as the major part of the load-carrying capacity is taken by self-weight alone).

Finally, some examples about modern problems related to vibration in bridges, showing that there is still much to learn:

–   The 300 metre long central box-girder steel span of the Rio-Niterói Bridge in Rio de Janeiro, Brazil, experienced in 1980 strong vertical oscillations due to vortex shedding excitation. The vibration was, however, not harmful to the safety of the bridge. This example shows that the vortex shedding phenomenon is not only restricted to suspension bridges alone.
–   In June 2000, only some few days after the opening, the London's Millennium Footbridge had to be closed after experiencing large lateral swaying motion excited

by walking people. It was found that the pedestrians synchronized their footsteps as the bridge swayed horizontally from side to side, because of the low lateral stiffness of the bridge deck – first very small deflections, but then increasingly larger as the motion of the footsteps tended to coincide with the swaying. A problem very much reminiscent of the collapse of the Angers Bridge in France back in 1850. What was it the sign said (and still says) on the Albert Bridge over Thames, also that in London: *"All troops must break step when marching over this bridge."*

Chapter 8

Peace River Bridge

Following the attack on Pearl Harbor in December 1941 a war program was initiated where a highway between the United States and Alaska, through Canada, immediately was to be built (facilitating chiefly the transport of military personnel and equipment, and, of course, oil). At Taylor, in between the cities of Dawson Creek and Fort St. John, approximately 500 kilometres to the northwest of Edmonton, the Peace River had to be bridged (Figs. 8.1 and 8.2). Earlier, the transport across the river was served during summer by a ferry, and during winter, when the river froze, by sleighs over the ice.

Fig. 8.1 North America, with British Columbia – the westernmost province of Canada – especially marked.

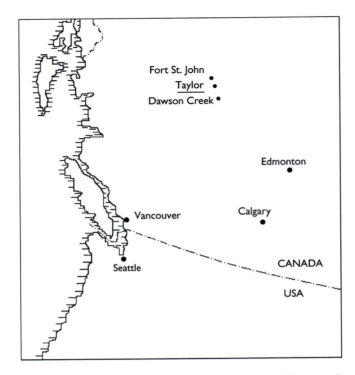

Fig. 8.2 The small community of Taylor – in between the cities of Dawson Creek and Fort St. John in Canada – where the Peace River Suspension Bridge was located.

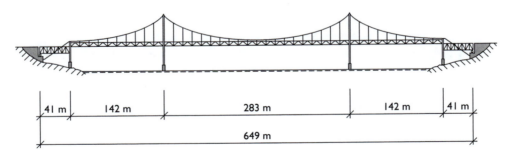

Fig. 8.3 Elevation of the Peace River Suspension Bridge at Taylor, Canada.

The construction of the bridge started in 1942, and on 30 August 1943 – only nine months after the construction began – the Peace River Bridge on the Alaska Highway was opened. With a main span of 283 metres and a total length of 649 metres it was the longest bridge on the Alaska Highway (Figs. 8.3 and 8.4).

With respect to the stability in winds the open stiffening trusses – 4.0 metre deep – provided a very stiff bridge (not to forget the Peace River Bridge was constructed only just short of three years after the Tacoma Bridge collapse). The stiffening trusses,

Fig. 8.4 A scenic view of the Peace River Bridge (photo taken 1950). Note the upstream sloping side of the tower piers, not only built because of increased stability in the rapid-flowing waters, but – most probably – primarily for the reduction of ice loads during winter. A nearby temporary timber bridge had in fact been damaged by ice pressure in 1942. (www.explorenorth.com/library/akhwy/ peaceriverbridge.html)

which supported the 7.3 metre wide concrete deck, were connected to the main cables by vertical 64 millimetre steel wire hangers. The main cable consisted of 24 individual parallel strands (bars), 48 millimetres in diameter, and separated apart from each other in an open-type solution (i.e. not compacted and enclosed) – a solution which makes the strands exposed to the weather conditions, but then, easier to inspect and maintain (Fig. 8.5).

The cable continues from the main span over the tower saddle (Fig. 8.6) into the suspended side span, where it is, at the end, splayed over a so-called cable bent (Fig. 8.7), which, besides supporting the main cable, at the same time is supporting the side span stiffening truss as well as the end span (Fig. 8.8).

For longer suspension bridges – where the cable force is high – the cable needs to be safely anchored into solid rock, but for shorter bridges, such as the Peace River Bridge, the pulling force was of such a low magnitude that the pure weight of the abutment (approximately 25,000 tons in this case) was enough to ensure sufficient resistance (Figs. 8.9 and 8.10).

The vertical lift exerted by the cables is resisted by the total weight of the abutment, and the horizontal pull is taken by friction of the footing underneath. The abutment itself should also balance the resultant lateral earth pressure of the soil (i.e. the difference between the active and passive pressure and the lever arm of the same).

The choice of a gravity anchor for the Peace River Bridge was not only based upon the fact that the pulling forces were low, but also that the rock was of a mudstone type (shale), of low strength and layered.

Fig. 8.5 The connection details of the vertical hanger; below to the stiffening truss and above to the main cable. (www.inventionfactory.com/history/RHAbridg/term/)

Fig. 8.6 The top of the towers, where the main cable passes over the saddle, transferring the load down vertically into the tower leg. (www.inventionfactory.com/history/RHAbridg/term/)

The first signs that the foundations of the bridge were defective came in 1955, twelve years after the bridge was taken into service, when it was discovered that one of the tower piers had been subjected to scour and had to be underpinned. But it was not until two years later that a more critical weakness of the foundations was shown.

Fig. 8.7 The splay saddle on top of the cable bent. (www.inventionfactory.com/history/RHAbridg/term/)

Fig. 8.8 From the cable bent down to the anchorage the strands are splayed, i.e. separated apart in order to provide necessary room for each strand to be safely and securely anchored to the gravity block. (www.inventionfactory.com/history/RHAbridg/term/)

Early on the morning of 16 October 1957, an unexpected bump in the roadway was reported from one of the drivers crossing the bridge. It was discovered that a water line close to the north abutment had sprung a leak, and was pouring out huge amounts of water in the ground – visible signs of ground movement at the abutment was also discovered. Possibly a small initial slip of the abutment – indicating the major collapse

Fig. 8.9 A safe and strong anchorage of all 24 individual strands of the main cable into the gravity block. (www.inventionfactory.com/history/RHAbridg/term/)

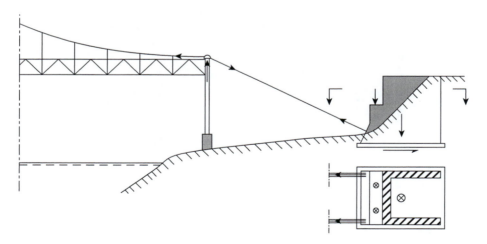

Fig. 8.10 The pulling forces from the two main cables, which are anchored in the abutment, are counterbalanced by the weight of the concrete and the soil and the reaction forces coming from the roller bearings supporting the end span (the end span is not shown in the figure above because it does not contribute to the horizontal stability).

Fig. 8.11 The collapsed Peace River Bridge. (http://collections.ic.gc.ca/north_peace/transport/ 02.07.html)

to come – had created the bump experienced by the drivers (as a misalignment of the deck either at the transition between end span and the suspended side span – i.e. at the cable bent – or possibly between the side span and the main span). The inspection personnel immediately realized the danger and closed the bridge to traffic. Just as the case was for the Tacoma Narrows Bridge the collapse of the Peace River Bridge came several hours later, so a large crowd of people could gather and witness the actual collapse. Suddenly, at approximately one o'clock in the afternoon, the north abutment slid towards the river some 3.5 metres and the end span and north suspended side span collapsed simultaneously (Fig. 8.11).

Whether the actual slide of the north abutment was because of a reduced capacity to resist the pulling force of the cables – because of the saturation of the ground – or that a landslide had been initiated, taking the abutment and the surrounding soil (or perhaps even the mudstone) with it, has not to this day been established. The actual collapse sequence was, however, as follows; as the abutment slid forward the cables slackened – transferring the self-weight of the end span to the end span itself (which deflected just as the end span of the Tacoma Narrows Bridge did – see Fig. 7.7 for comparison – however, this time not because of the lack of counterbalancing weight from a collapsed main span). The cable bent became eccentrically loaded and deformed heavily under the combined action of bending and axial normal force, which made it buckle and break (Fig. 8.12).

The end span could of course also have helped to push the cable bent forward – even though the end span is not shown in the figure above – as soon as it came in contact with the front wall of the abutment that is. The broken and deformed cable bent can be seen lying on the ground to the left in figure 8.11, leaning backwards supported by its plinth (and behind it the collapsed end span). In the same figure the deflected main span can be seen – it oscillated heavily at the time of the collapse, but was not

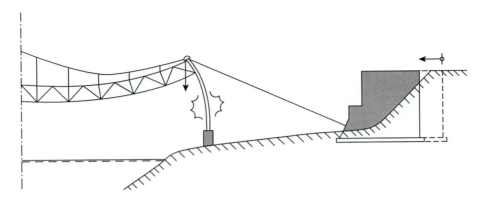

Fig. 8.12 The deformation of the cable bent as the abutment slid.

destroyed (opposite to the Tacoma Narrows Bridge, where the main span was destroyed while the end spans remained intact). Due to the slackening of the main cable – now straightened in between the tower top and the anchorage at the abutment – and the uneven loading on either side the north tower deformed in just the same manner as the cable bent, however without failing, thus saving the main span and the south side span and end span.

There were suggestions that the slide of the abutment had been triggered by the frequent use of the bridge by heavy vehicles, far exceeding the weight limit of the bridge, however, at the time of the collapse the bridge was closed, giving a minimum of pulling force in the cables. And the fact that the bridge was closed well in time of the collapse also resulted in that there were no casualties.

After the partially destroyed bridge was dismantled – a process just as complicated and time-consuming as the erection, perhaps even more one could imagine – the Peace River Suspension Bridge was replaced in 1960 by a continuous truss bridge in the very same location, and it still remains in use today on the Alaska Highway.

Chapter 9

Second Narrows Bridge

On 17 June 1958, eight months after the collapse of the Peace River Bridge (see Chapter 8), and while the same was being dismantled, yet another major bridge in British Columbia collapsed, however, this time a bridge under construction. In Vancouver, over the Burrard Inlet at the second narrows – between the Strait of Georgia at the Pacific Ocean and the mouth of the Fraser River – the construction of a new highway bridge had begun in November 1957 (Fig. 9.1).

The Second Narrows Bridge was built alongside an existing old bridge for combined railway and highway traffic built in 1925 (the *first* Second Narrows Bridge).

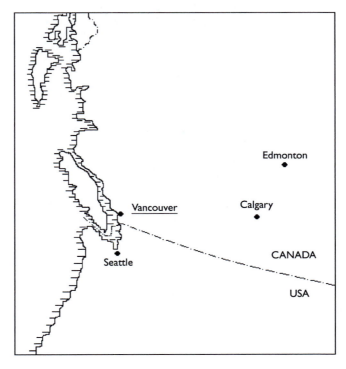

Fig. 9.1 The Second Narrows Bridge is located in Vancouver in British Columbia, Canada (see also Figs. 8.1 and 8.2).

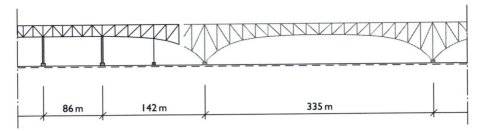

Fig. 9.2 Elevation of the Second Narrows Bridge under construction.

Fig. 9.3 The temporary truss (i.e. the falsework) supporting the north anchor span during construction. (McGuire: Steel Structures. With kind permission of Pearson Education Publications).

This existing bridge had been hit by ships on numerous occasions, and it was therefore decided that the second bridge – constructed for highway traffic only (having six lanes) – should be a high-level bridge. It was constructed as a continuous truss bridge (in all 1292 metre long), with the main 335 metre navigation span as a cantilever construction having two anchor spans (142 m each) on either side (Fig. 9.2).

At the time of the collapse the north anchor span was being erected, and in order to reduce the maximum cantilever length a centrally located falsework support was being used. In contrast to the anchor spans of the Quebec Bridge back in 1907 – which were constructed using an all-span length falsework (see Fig. 4.6) – it was in this case decided to limit the amount of extra supports to just one because of the water (Fig. 9.3).

Fig. 9.4 The collapse of the Second Narrows Bridge. (McGuire: Steel Structures. With kind permission of Pearson Education Publications).

On the afternoon of 17 June 1958, despite being supported, the north anchor span collapsed without any warning through the failure of the temporary truss (Fig. 9.4).

When the anchor span fell the adjacent span also collapsed, as it lost its support when the pier was pulled away. A total of 18 workmen and engineers were killed. A Royal Commission of Inquiry investigated the cause of the collapse and soon found the answer. It was not the buckling capacity of the temporary truss that was inadequate, as could have been expected; instead it was found that the lower transverse beam at the bottom of the falsework truss had failed. The purpose of this beam was to distribute the concentrated load – coming from the vertical members of the truss – in order to spread it over a larger area on top of the piles. The profile – or profiles, for there were probably more than one, lying parallel close to each other as a grillage – was a wide-flanged rolled standard steel girder (36 WF 160) (Fig. 9.5).

As the concentrated load is transferred directly through the web to the support there is no risk of web crippling (i.e. local buckling of the upper part of the web) as no shear force is present (i.e. the girder is not subjected to bending), instead the instability phenomenon to consider in design should be global buckling of the entire web – an instability phenomenon that is more or less similar to the Euler buckling case where both ends of a strut are fixed (Fig. 9.6).

If the upper flange, though, is not fixed in the transverse direction, the effective length factor is somewhere in between fixed end conditions ($k = 0.5$) and that of a free-swaying cantilever ($k = 2.0$), which have the result that the capacity to carry concentrated loads is lowered even further (Fig. 9.7).

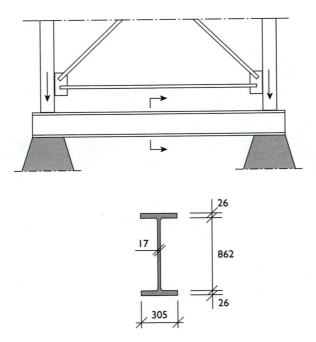

Fig. 9.5 Cross-section of the transverse girder.

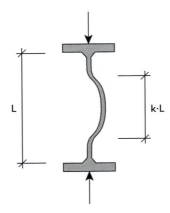

Fig. 9.6 Overall buckling of the entire web when the load is transferred directly through the web (with a column effective length factor k equal to 0.5).

The exact conditions for the transverse girder(s) is difficult to ascertain in detail, but nevertheless, the capacity was far from being enough as the huge cantilever up above lost its temporary support and plunged into the water below (Fig. 9.8).

Swan, Wooster & Partners – the designer of the bridge – shamefully blamed the error in design on a young and inexperienced engineer, who, as a matter of fact, was among the casualties. This engineer was supposed to have used the flange thickness (26 mm)

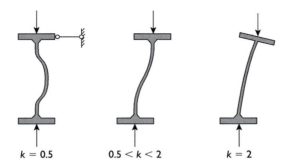

k = 0.5   0.5 < k < 2   k = 2

Fig. 9.7 The effective length factor *k* with respect to web buckling for different boundary conditions.

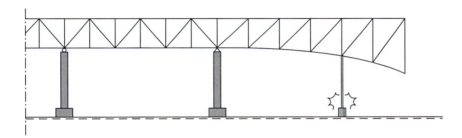

Fig. 9.8 When the web of the temporary transverse girder below buckled the entire weight of the cantilever above was transferred to the truss itself – which it had not been designed for – so it broke in bending and collapsed (see also Fig. 9.4).

instead of the web thickness (17 mm) in computing the resistance of the transverse girder to concentrated loads, but even if this is true there are two remarks to be made. First, that there, as a rule, *never* should be relied upon an unstiffened web to carry a concentrated load, whatever the real capacity is, and secondly, that any computation made by an inexperienced engineer at a company should, also as a rule, be checked (and double-checked) by his supervisor (a senior engineer at the office). The company should then stand united and not blame one of their employees if a mistake is made!

A couple of two small web stiffeners at each end of the transverse girder should have saved the bridge (stabilizing the web and the upper flange in these locations (Fig. 9.9).

In figure 9.3 it is indicated that vertical web stiffeners *actually were* used, however, not directly under the load application points (and these indications could also very well be just bolting, holding the grillage tight together).

And finally, to summarize what really happened: too much of the concentration was focused on the construction up above of the cantilever truss, that a proper design of a somewhat negligible, but for the load-carrying capacity ever so crucial transverse girder down below was neglected.

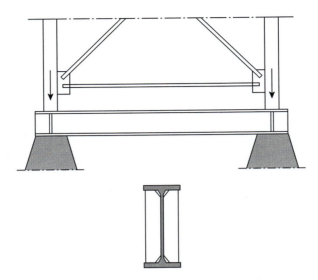

Fig. 9.9 Vertical web stiffeners at the load application points would have saved the bridge, but were unfortunately missing.

Two years later the construction was, however, completed, and on 25 August 1960, the bridge was officially opened. In order to commemorate the lives of the workmen that were lost in the collapse in 1958 (and also to those who completed the bridge, one could imagine), the name of the bridge was changed to Ironworkers Memorial Second Narrows Crossing in 1994.

Chapter 10

Kings Bridge

In September 1957 the construction of a multi-span high-level highway bridge in Melbourne, Australia (Fig. 10.1), named the Kings Bridge, started.

The bridge crosses over the Yarra River as well as over a railway and a couple of streets, in a north-southerly direction. Due to poor soil conditions it was decided upon using high-strength steel in order to keep the self-weight down of the I-girder bridge (Fig. 10.2).

In April 1961 – three and a half years after the construction began – the bridge was taken into service. Fifteen months later, on 10 July 1962, at 11 o'clock on a cold

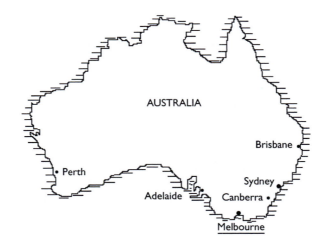

Fig. 10.1 The Kings Bridge was located in Melbourne in the southernmost part of Australia.

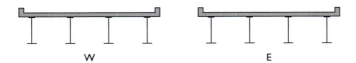

Fig. 10.2 The bridge consisted of a large number of spans – having two separate carriageways – each built up by a concrete deck supported by four parallel I-girders, in composite action with the deck through shear connectors at the top flange.

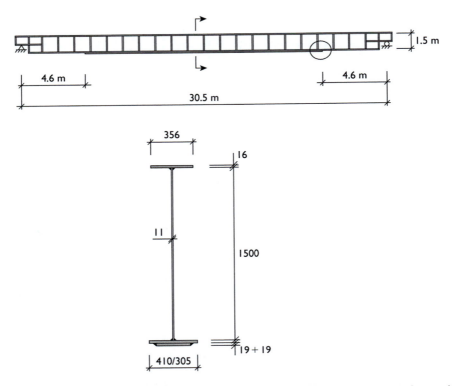

Fig. 10.3 Elevation and cross-section of the *outer* girders of the second span on the south end of the bridge. The inner girders had 356 millimetre wide cover plates, which were terminated 5.3 metres (instead of 4.6 metres) from the ends.

winter's morning, as a 45-ton heavy vehicle was passing over the second span on the western carriageway – having approached the bridge from the south – this span broke and collapsed. The span was 30.5 metre long and the I-girders were reinforced by extra cover plates attached to the tension flange in the maximum-moment region (Fig. 10.3).

The purpose of this extra cover plate was three-folded:

–   To increase the bending moment capacity,
–   To lower the neutral axis and thereby lower the tensile stresses in the bottom flange, with respect to fatigue a good initiative,
–   To reduce the necessary preheating temperature (for the longitudinal flange-to-web fillet weld), instead of choosing a 38 millimetre thick flange plate that is.

In order to achieve a smooth transition – rather than an abrupt change in stiffness – the cover plates were tapered (i.e. gradually reduced in width) at the ends (the south end encircled in Fig. 10.3) (Fig. 10.4).

When the lorry passed over the second span on the western carriageway the position was in between girders number one and two (Fig. 10.5).

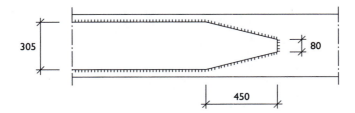

Fig. 10.4 The cover plates were tapered over a length of 450 millimetres. Fillet weld dimension: 5 mm – automatic welding for the parallel sides, and manual welding for the tapering and the 80 mm transverse end.

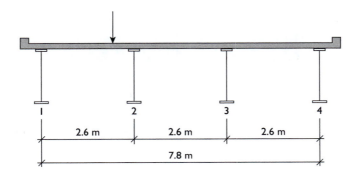

Fig. 10.5 Cross-section of the second span on the south end of the Kings Bridge, and the approximate position of the lorry that initiated the collapse.

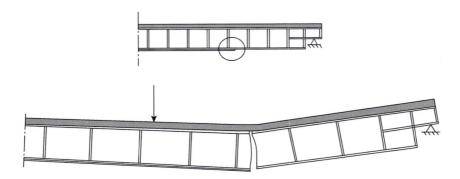

Fig. 10.6 The broken girders.

It was found that all four girders had fractured in a brittle manner in the position of the cover plate ends (Figs. 10.6 and 10.7).

In three positions the crack was complete, i.e. had propagated across the entire depth of the girder (the south ends of girder No. 1, 2 and 3) and in four positions the crack had arrested in the web below the upper flange (the south end of girder No. 4, and the north ends of girder No. 2, 3 and 4) (Fig. 10.8).

Fig. 10.7 Complete fracture of I-girder No. 1 on the western carriageway. The photo is taken inside the bridge looking north, in between girder No. 1 and 2. (Hopkins: A span of bridges – an illustrated history).

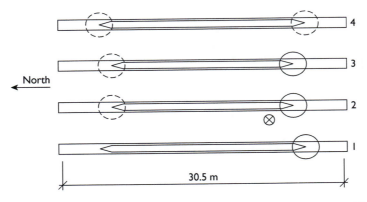

Fig. 10.8 Complete fracture in the south ends of girder No. 1, 2 and 3, and partial fracture of the south end of girder No. 4 and of the north ends of girder No. 2, 3 and 4. For the sake of simplicity the flanges have been turned upside down – the flange widths have also been enlarged (in relation to the span length).

The bridge was immediately closed to traffic and there were no casualties, and even the lorry – which caused the collapse – safely reached the next span. In fact, the collapse of the span was not complete because of some screening walls underneath the bridge, and they were supporting the sagging concrete deck, to which, in its turn, the steel

Fig. 10.9 The stress concentration effect at the weld toe.

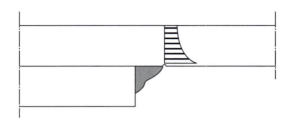

Fig. 10.10 Additional stress raising effects coming from the roughness of the surface, and, especially, melting ditches.

girders were hanging. The western screening wall can actually be seen in Fig. 10.7, just outside girder No. 1.

Before coming to the findings of the Royal Commission of Inquiry we will shortly discuss why this cover plate end is extra sensitive with respect to cracking. First and foremost the weld and the heat-affected zone in the parent metal (surrounding the weld) is a region which is weakened with respect to fracture toughness, as it is containing impurities, having a large concentration of carbon, small micro-cracks being present and having coarse grains. Still being tapered off, the cover plate end is also an inevitable stress raiser due to the stress concentration effect (Fig. 10.9).

The roughness of the surface of the manual fillet weld, as well as melting ditches will increase this local stress raising effect even further (Fig. 10.10).

When the weld cools, and the contraction is restrained by the surrounding cold metal, small micro-cracks will form (as well as the forming of residual tensile stresses – counterbalanced by residual compressive stresses in the surrounding metal). One way of limiting the formation of these cracks (and the residual stresses) is to preheat the metal pieces to be joined; however, regions which still are especially prone to crack are corners, because of the combined effect of the start/stop-procedure of the welding and the restrained contraction thereafter during cooling (Fig. 10.11).

All these effects taken together make this region in the tension flange – at the cover plate end – very probable for the initiation and the following propagation of a *slow* fatigue crack (a small incremental increase in crack length for each new loading cycle) or a *fast* brittle fracture (the final failure starting from an existing crack for just one additional loading cycle). The crack will propagate *perpendicular* to the applied normal tensile stress, starting from the weld toe in the heat-affected zone (Fig. 10.12).

When the cracked girders were examined closer in detail after the collapse, it was found that all the fractured locations had *existing cracks* originating already from when the girders were manufactured or close in time thereafter. The evidence for this was because of the crack surfaces were either covered with paint or being corroded. These

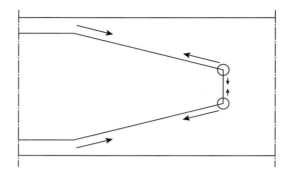

Fig. 10.11 The corner regions, where the probability for crack initiation is the highest. Here the combined effect of residual tensile stresses in the transverse and longitudinal direction is the highest.

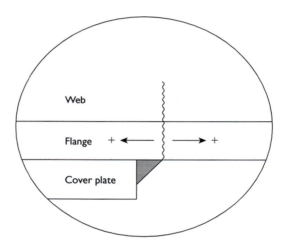

Fig. 10.12 A crack will initiate from the weld toe, in the heat-affected zone, and then propagate upwards into the web, perpendicular to the applied normal tensile stress.

existing cracks had in general a width equal to the width of the cover plate end (i.e. 80 mm) and extended as far as through the flange plate and 25 millimetres up into the web. In one case though, at the south end of girder No. 2, the crack had propagated a long distance up into the web (1100 mm), as well as having fractured the entire lower flange – thus having a load-carrying capacity being more or less completely gone already from the start (or perhaps more probable, some time during the first winter, that is, in 1961). At the time of the failure of the span, one year later, the three remaining intact girders were not able to carry the heavy load so they broke (at the same time as the net-section of girder No. 2's south end also broke). The combined effect of a heavy vehicle, impact loading, existing cracks and low temperature (it was +7°C that morning, and +2°C during the night) made the girders break. The Royal commission of Inquiry found numerous other contributing factors (besides discovering the presence of existing cracks). First of all it was established through chemical analysis that the high-strength steel was not suited for welding due to its high carbon and manganese content. The carbon equivalent value was as high as 0.61, and would require a careful pre-heating procedure of the steel before being joined by welding. The steel had been pre-heated, but not at all to such a level that is required given the high carbon equivalent value. With respect to modern standards steels having a carbon equivalent above 0.6 are normally considered unweldable (as it requires a pre-heat temperature of 100–400°C). The chemical analysis also revealed that the content of nitrogen was high, indicating a steel susceptible to ageing (i.e. the embrittlement with time).

The welding sequence was also criticized – by welding the cover plate sides in the wrong order (starting with the longitudinal welds and finishing off with the transverse welds at the ends) too much constraint was built in, subjecting the welds to unnecessary strain. However, it was not the welding sequence – and the in-built stresses – that was the major criticism with respect to welding coming from the Commission. It was found that the electrodes had not been sufficiently dried in order to avoid moisture being absorbed by the weld, thus developing cold cracking in the heat-affected zone at the weld toe (see Fig. 10.12) due to the hydrogen content. This hydrogen-induced cracking – especially at the cover plate ends – together with the brittle material, explained why the cracks had formed already when the girders were being produced. When the other bridge spans were investigated it was found that a total of 86 transverse welds at cover plate ends were having these toe cracks, and the Commission pointed out the complete lack of proper inspection routines that would have discovered these critical faults. Perhaps had the screening walls enclosed the bridge in such a manner that regular inspection was made impossible? It was anyway in line with the experience and competence of the design team – the lead designer had not constructed a bridge earlier in his career, and the rest of the team had previously not dealt with high-strength steel in any type of construction. In the decision to choose high-strength steel (a yield strength of 325 MPa for the flange plates) they saw the opportunity to not only save weight, but also to reduce costs (less steel and less weight to be carried by the substructure), however, without having the proper experience in how to produce welded I-girders in such a material. The allowable stress level for static loading was not only high, but the fatigue strength used was also higher than what was normal for mild steel. Of course a high-strength base material requires a higher stress to fracture in cyclic loading, but this is counterbalanced by the presence of a higher residual stress level and a higher probability of existing cracks in a welded construction. So the I-girders were overutilized with respect to fatigue loading, which was also pointed out by the Commission.

If we consider the girder depth in relation to the span length we could also see that this relation is very small (approximately 1:20), which strongly indicates a very flexible bridge, i.e. having large deflections when being loaded. The primary bending stresses as such do not increase because of this, but there are secondary effects that perhaps were not considered. At large curvatures there is a phenomenon called flange curling, a visible phenomenon in wide-flanged I-girders with small thicknesses, however, still being present also in thick flanges. As the deflection of the I-girder takes place there is a tendency for the flanges to bend in the transverse direction towards the neutral axis (only the lower flange in the case of the Kings Bridge, as the upper flange was connected to the concrete deck). Due to the curvature the tensile stresses pull the lower flange upwards (and the upper flange is by the same action pushed down because of the compression, if it would have been free to do so that is). This secondary effect therefore introduces additional bending stresses (in the transverse direction) to the lower flange. Another effect that was not considered was the shrinkage of the concrete, which was introducing an additional tensile stress to the lower flange in the neighbourhood of 15 MPa (as was found in an analysis made during the work of the Commission). The creep of the concrete (i.e. its gradual loss of stiffness with time when being subjected to loading) was most certainly also not considered, which increases the stresses in the steel even further.

The investigation also showed that the steel did not contain the desired ductility required by the designer. At the time of the construction of the Kings Bridge these requirements were omitted because of the inability of the steel industry to produce high-strength steel in such quantities, having a certain minimum notch ductility at low temperatures. However, if these requirements would have been provided for, the notch ductility would still have been inadequate, as they were far from meeting the standards. Charpy impact notch toughness tests performed by the Commission showed that the steel – as it was delivered – had a value as low as 3 J at 0°C (well below the minimum requirement of at least 20 J at the lowest operating temperature). The minimum required notch ductility is to ensure that the structure responds in a fracture tough manner (even at low temperatures), especially in regions where the stress concentration is high. The action should be such that the local stresses at a notch should be able to reach yielding without there prior being a brittle fracture – this is the behaviour desired by every bridge designer (as well as the owners). This difference in response, between slow crack propagation (read: fatigue) due to yielding at a notch (e.g. a crack tip) and fast crack propagation (read: brittle fracture) due to cleavage of the atom bonds, is the background for the ever so vital robustness requirement of a bridge (i.e. to withstand excessive loading, damages and local cracking).

We could perform a simple fracture mechanics analysis by using the results from the impact notch tests (by assuming that plain-strain linear elastic fracture mechanics is applicable, which it is). The stress intensity factor, K, for a centrally located crack in a wide plate subjected to uniform loading is (Eq. 10.1):

$$K = \sigma \cdot \sqrt{\pi \cdot a} \cdot f \tag{10.1}$$

where:
σ applied nominal stress,
a crack length,
f factor taking crack length and geometry into account.

We assume a crack all along the length of the cover plate end (having a length equal to $2a = 80$ mm) (Fig. 10.13).

By using the results from the Charpy V-notch impact tests we find the fracture toughness, K_c, of the flange (i.e. its ability to withstand the presence of this crack without fracturing at a certain given loading) from the following correlation (Eqs. 10.2 and 10.3):

$$K_c = 11.76 \cdot (CVN)^{0.66} = 11.76 \cdot (3)^{0.66} = 24.3 \, \text{MPa}\sqrt{\text{m}} \tag{10.2}$$

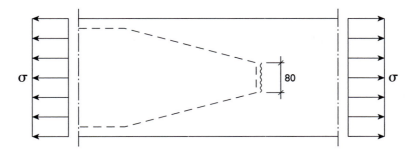

Fig. 10.13 A centrally located crack all along the cover plate end.

$$2a/W = 80/410 = 0.20 \Rightarrow f \approx 1.0 \tag{10.3}$$

Finally, we find the actual stress that would fracture this flange plate at the presence of the assumed crack (Eq. 10.4):

$$\sigma = \frac{K_c}{f \cdot \sqrt{\pi \cdot a}} = \frac{24.3}{1.0 \cdot \sqrt{\pi \cdot 40 \cdot 10^{-3}}} = 68.5 \, \text{MPa} \tag{10.4}$$

The magnitude of this stress is so low that we can conclude – given the high allowable stress level – that the bridge was doomed to fail.

The faulty design of the Kings Bridge was as if all the lessons learnt from the failures of the welded bridges in Germany and Belgium in the late 1930's (see Chapter 5) was completely forgotten. In fact, the bridge at Rüdersdorf also failed due to transverse welding in a tension flange, so the mistake was here repeated. And finally, to sum it all up, in the words of the Commission:

"We conclude that the primary brittle fractures in these girders were due to the brittle nature of the steel. They were 'triggered' by the cracks present at the toes of the transverse weld at the cover plate ends."

Chapter 11

Point Pleasant Bridge

At five o'clock in the afternoon on 15 December 1967, in the middle of rush-hour traffic, the highway bridge over the Ohio River at Point Pleasant (Fig. 11.1) suddenly and quite unexpectedly collapsed, killing 46 people.

The Point Pleasant Bridge (also called the "Silver Bridge" because of the aluminium paint) was a 445 metre long suspension bridge connecting the shores of West Virginia and Ohio (Figs. 11.2 and 11.3).

The bridge was constructed in 1928 and had some unique and special features, among them the most important one being that the main cable – designed as an eye-bar chain – was integrated with the stiffening truss (Figs. 11.3–11.6).

Before the age of steel wires the main cables in suspension bridges were built up of wrought-iron eye-bar chain elements, e.g. like it originally was in the Menai Strait Suspension Bridge in Wales from 1826 (see Fig. 11.5) – today this bridge has been

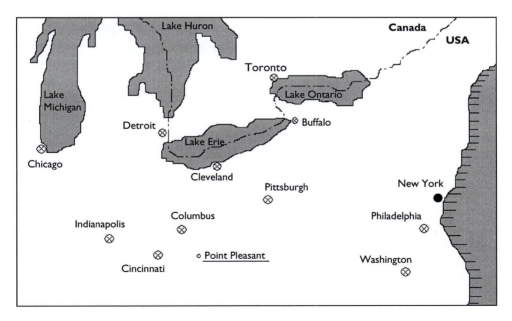

Fig. 11.1 The City of Point Pleasant is located in West Virginia, USA, and the Ohio River forms the border between West Virginia and Ohio at this location.

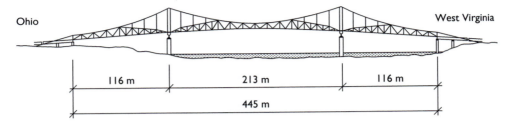

Fig. 11.2 Elevation of the Point Pleasant Bridge over the Ohio River. (Dicker: Point Pleasant Bridge collapse mechanism analyzed. Civil Engineering, ASCE)

Fig. 11.3 The Point Pleasant Bridge (in the foreground). (http://wikipedia.org/wiki/Silver Bridge)

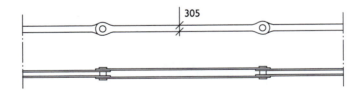

Fig. 11.4 The chain suspension cable of the Point Pleasant Bridge – built up by two 51 × 305 mm² eye-bar steel elements, and linked together by pins – just like a bicycle chain.

strengthened though using steel bars. When the magnificent Brooklyn Bridge over the East River in New York was built in 1883, and steel wires came to be used for the first time in bridge engineering, eye-bar chains became more or less obsolete thereafter (besides repair and strengthening of old existing suspension bridges of course). However, parallel to the development of high-strength steel wires, eye-bars (in steel) were still used now and then. Due to the unusual choice of configuration of the Point Pleasant Bridge, where the main cable was integrated in the stiffening truss (as being the upper chord), the cable needed to be rigid and stiff as it became subjected to

Fig. 11.5 The eye-bar chain suspension bridge at Menai Strait, Wales, similar in design to that of the Point Pleasant Bridge. (www.anglesey.info/Menai%20Bridges.htm, with kind permission of Phil Evans)

Fig. 11.6 A simplified model of the Point Pleasant Bridge main span.

compression in the centre part of the span (ruling out a wire solution). This particular bridge type was named after its inventor, the American engineer David B. Steinman. He designed and built the world's first ever eye-bar chain bridge, with the cable integrated with the stiffening truss (the Steinman Type), in Brazil in 1926. Steinman later became known for his prediction that the Tacoma Narrows Bridge would fail, after that his own design proposal was rejected for the much slender solution that was finally chosen (see Chapter 7). In 1957 he crowned his career when he designed the mighty Mackinac Bridge – a suspension bridge (with a main span of 1158 m) in the northern region at Lake Michigan, USA.

As a simplified model of the Point Pleasant Bridge main span one can regard the system as a simply supported truss suspended at both ends (Fig. 11.6).

The choice of using eye-bar chains also made it necessary to use specially devised "rocker towers", as the main cable was unable to slide over the tower saddle (as normal suspension bridge cables do). To accommodate for difference in loading between the spans and for the temperature deformations, the towers were made able to lean in the

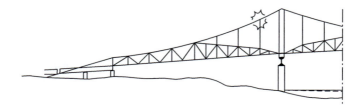

Fig. 11.7 The location of the fractured eye-bar chain joint.

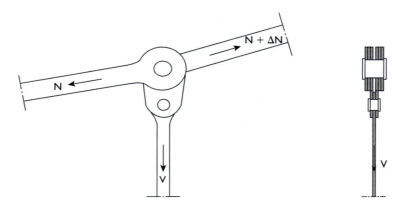

Fig. 11.8 Static equilibrium in the upper hanger joint.

longitudinal direction of the bridge through the application of "hinges" at the base of the towers just below the deck (see Fig. 11.2).

Despite being unusual in design the bridge served its purpose well for almost 40 years, but on that fatal afternoon in December 1967 it was no longer capable of carrying the service load it was intended for. First the side-span on the Ohio side fell into the river, and then the main span and the other side-span followed almost instantaneously, and all cars on the bridge plunged down as well. After the first couple of days of chaos an investigation into the cause of the collapse commenced, where as many parts as possible of the bridge superstructure was salvaged from the river bed. A large number of probable causes were first considered (among them aerodynamic instability and sabotage), however, these were eliminated when a fractured eye-bar was found. The broken eye-bar was identified with respect to its original location, and it was found that it had been positioned in the second panel counting from the Ohio tower on the upstream north side, with its fractured end connected to the eye-bar in the first panel (Fig. 11.7).

In each chain joint a vertical hanger is attached, which is carrying a vertical reaction force, V, coming from the self-weight of the stiffening truss and deck and from traffic load upon the bridge. Vertical equilibrium in the joint is met by an increase in the normal force, N, in the eye-bar chain element closest to the towers (Fig. 11.8).

A computation of the actual stresses at the time of the collapse did not show that the dimensions of the bridge members were too small – the bridge had been able

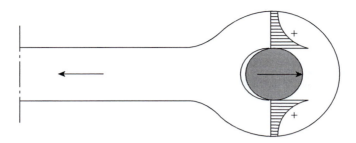

Fig. 11.9 Stress concentration effects in the eye-bar end.

to carry the gradual increase in loading (from 1928 to 1967) without exceeding the allowable stresses. However, it was soon realized that the actual loading of the eye-bar hangers was high, as they were made out of high-strength steel (the higher the strength the higher the allowable loading). For the eye-bar shank it was not a problem, however, the tensile stresses in the end parts – taking the stress concentration effects into account – were found to be very high (Fig. 11.9).

For static loading (read: dead load) these large localized tensile stresses at the edge of the pin hole is not a problem. Local yielding can not be avoided in steel structures, as it will occur at first heavy loading, but favourable residual compressive stresses will always, as a result of the early yielding, be formed in these regions. However, when the bridge is subjected to repeated loading (read: traffic) over a long period of time, fatigue cracks can eventually occur in locations where tensile stresses are high (despite the presence of residual compressive stresses). An eye-bar chain (or a wire cable for that matter) is normally not susceptible to the initiation of fatigue cracking as the number of repeated loadings is small (read: due to the fact that the influence area is large) – a continuous flow of vehicles over the bridge will not markedly alter the stress in the cable, and hence not the fluctuation in stress as well. For the vertical hangers though, each passing vehicle close to a hanger will produce large variation in stress, which is repeated for each new passage. However, there are more factors to consider that explain why it was the eye-bar chain that failed instead of any of the vertical hangers.

As the load from the traffic is transferred by the vertical hangers to the eye-bar chain it is followed by a rotation of the chain joints (and by a deflection down of the joint), as they are adjusting themselves to the new loading position and loading level (see also Fig. 7.10) (Fig. 11.10).

This rotation creates fretting due to the sliding friction between the pin and the hole edge, and at the ends of the contact zone it adds to the tensile stresses from the normal force, which drastically increases the probability of a fatigue crack being initiated at the surface (especially as the number of loading cycles equals that of the hanger). When the hanger is unloaded (with respect to the service load) the eye-bar chain elements rotate back, now creating sliding friction in the opposite direction (Figs. 11.11 and 11.12).

In an almost 40-year-old structure, where the maintenance was said to have been poor, there was also the possibility that adjacent eye-bar plate surfaces in the upper hanger joints (see also Fig. 11.8) had rusted stuck. If this was the case then rotation

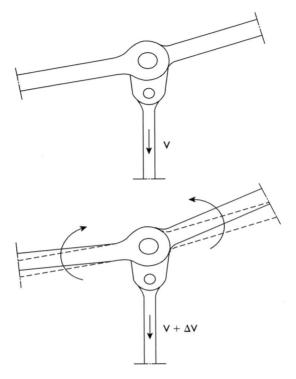

Fig. 11.10 Rotation of the joint as the load, transferred by the vertical hanger, is increased.

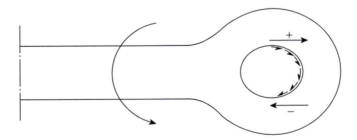

Fig. 11.11 Sliding friction between the pin and the hole edge as the eye-bar chain element is rotating back to its original position. Due to the restraint a partial fixed end moment is introduced.

of the eye-bar elements would have been prevented, introducing secondary bending stresses into the joint (as well as to the eye-bars) as soon as traffic load was carried by the hanger (Fig. 11.13).

Consider the following similitude, where two eye-bar elements – in the upper case – are completely free to rotate and adjust to the applied load, and – in the lower

Fig. 11.12 At the end of the contact zone, on the upper side (as well as on the lower side), a fatigue crack could be initiated due to the combined effect of fretting and normal tensile stresses (see also Fig. 11.9).

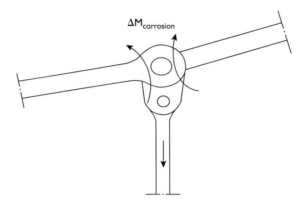

Fig. 11.13 Secondary bending stresses are introduced if the joint has rusted stuck.

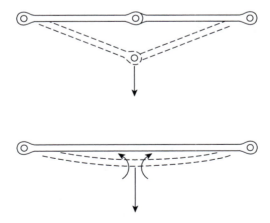

Fig. 11.14 The upper case where the eye-bar joint is free to rotate, and the lower extreme case where there is no rotational capacity at all, and therefore secondary bending stresses are being introduced.

case – where there is no rotational capacity at all left in the centre joint (and that is why it has been omitted) (Fig. 11.14).

The negative effect of corrosion in the eye-bar hanger joints due to lack of proper inspection routines and maintenance procedures can best be summarized in the words of Daniel Dicker 1971:

> "It seems somewhat ironic that when dealing with moveable bridges, such as bascule bridges, designers are well aware that special attention must be paid to pin connections, and that had positive lubrication been provided at the pins of the Point Pleasant Bridge so that necessary rotations of the members could occur easily and without the development of the large friction, the disaster, in all likelihood, would not have occurred."

He was mainly considering the sliding friction in between the pin and the hole edge, but his conclusion also applies to friction and corrosion problems of pinned joints in general – and as everybody knows, a bicycle chain also needs to be attended to every once in a while by applying lubrication in order to operate smoothly as it should.

In many different analyses and reports about the bridge collapse the effect of corrosion is often mentioned, but then only as an accelerating factor for the initiation and propagation of fatigue cracks – different phenomena such as *stress corrosion cracking* (for the initiation of small cracks in high-strength steels under the influence of tensile stresses and a corrosive environment) and *corrosion fatigue* (i.e. the propagation of an existing crack in a corrosive environment) have been discussed. Exactly how corrosion played a part in the collapse is probably never to be determined in detail, but it was most definitely an important and contributing factor.

The most probable failure sequence was as follows; a fatigue crack in the upper portion of the eye-bar end (see Fig. 11.12) continued to grow upwards, perpendicular to the applied tensile stress (see Fig. 11.9), for each new loading cycle (of heavy vehicles passing the bridge) – gradually increasing the stress concentration effect (and hence the crack propagation rate) – until, finally, it reached a certain critical length and net-section (brittle) failure of the upper half occurred. The entire load was then carried by the lower portion, which deformed rapidly under the combined action of tension and bending, and broke more or less immediately in a brittle manner (Fig. 11.15).

As all the loading was then transferred to the remaining and intact eye-bar this had to carry twice the load as before, and above all, being now eccentrically loaded. It did not take that many additional loading cycles before it had worked itself free from the pin, resulting in the complete separation of the chain link, and the eventual collapse of the entire bridge. There were many other highly stressed locations though on the eye-bar chain link that could have preceded the failure of this particular member, but the combined effect of high repeated loading (not necessarily the highest), corrosion that both increased the sliding friction between hole edge and pin, and that prevented rotation of the joint that produced secondary bending stresses, and above all, the existence of a flaw in the material at the region where the stress concentration effect was the highest – the last parameter being the most important one in explaining why there is such a high uncertainty to pinpoint an exact crack location point. But then, the failure *had* to come at the north side of the bridge as only two out of three possible lanes were in use – the south side lane was transferred into a sidewalk, so the traffic

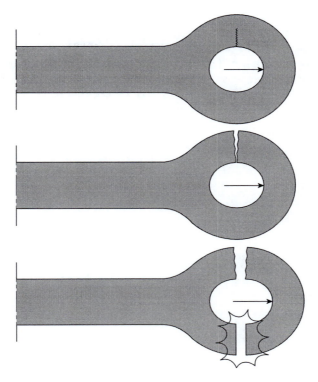

Fig. 11.15 A probable failure sequence.

was centred towards the north side of the bridge, and thus the cable there carried a larger proportion of the load.

Using only two parallel eye-bars – having a thickness of 51 millimetres each and with high allowable stresses – was, in hindsight, not a good choice. First of all the system is not structurally redundant – as soon as an eye-bar is lost it is impossible for the eccentrically positioned remaining one to carry the load alone. In a system which consists of a larger number of parallel eye-bars (and of more modest thicknesses) – that was the common case in old suspension bridges during the 1800's (see e.g. the Menai Bridge in Fig. 11.5) – the structural redundancy becomes markedly increased. Secondly, the thickness of the Point Pleasant Bridge eye-bars made them susceptible to brittle fracture – despite being heat-treated as they were. Structural members subjected to cyclic tension with high allowable stresses should, as a rule, be kept down in thicknesses – the toughness (and not only the redundancy) would definitely have been favoured by choosing a larger number of eye-bar plates of more modest thicknesses. Thicker plates have a reduced capacity to respond in a ductile (fracture tough) manner at the presence of a notch (flaws, fatigue cracks, scratch marks or other damages to the surface) – see also Chapter 5. The same mistake was made here – regarding the choice of using thick members instead of several thinner – as for the welded bridges in Germany and Belgium in the late 1930's (and early 1940) that failed.

Following the collapse much attention was focused on bridge inspection routines and maintenance procedures, and two bridges similar in type to the Point Pleasant Bridge were immediately closed to traffic; the St. Marys Bridge over the Ohio River in West Virginia – upstream of Point Pleasant Bridge – and Hercilio Luz Bridge in Brazil, the bridge built by Steinman in 1926. The St. Marys Bridge was demolished in 1971, and the Hercilio Luz Bridge was transformed into a pedestrian bridge. In 1969, a new highway bridge replaced the Point Pleasant Bridge, located one and a half kilometres downstream. This new bridge is a cantilever truss bridge, and was named the Silver Memorial Bridge.

Chapter 12

Fourth Danube Bridge

On the evening of 6 November 1969, there was some unexpected buckling occurring during the final stages of construction of the Fourth Danube Bridge in Vienna, Austria (Fig. 12.1).

Far away from the actual bridge site one could hear, with an interval of five seconds, three loud bangs, almost like explosions, and they all came from the bridge that had buckled under excessive loading. As the bridge not yet had been taken into service, the loading could only be coming from self-weight, which made it quite mysterious.

The Danube Bridge was a continuous box-girder bridge in three spans, having haunches over the inner supports. The total bridge length was 412 metres, with a main span of 210 metres (Fig. 12.2).

The bridge was a twin-box girder bridge, having a 5.2 metre deep and 32.0 metre wide cross-section in mid-span. The top and bottom flanges were stiffened in the longitudinal direction with simple flat plates (Fig. 12.3).

In order to minimize the disturbance for the boat traffic on the River Danube, one had chosen the free cantilevering erection method. By successively assemble the box sections to the ends of the cantilevers – instead of using a supporting falsework – there was a high demand for proper anchorage at the end supports, as well as a sufficient

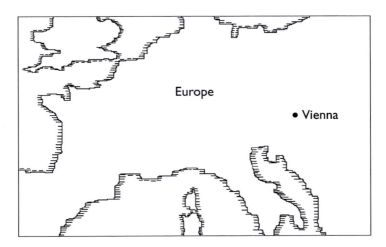

Fig. 12.1 The Fourth Danube Bridge was located in Vienna in Austria, Europe.

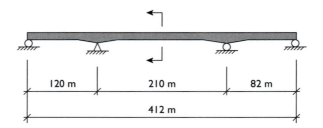

Fig. 12.2 Elevation of the Fourth Danube Bridge.

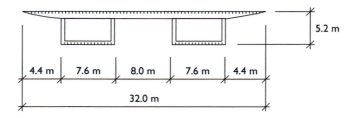

Fig. 12.3 Cross-section of the Fourth Danube Bridge.

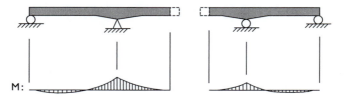

Fig. 12.4 Moment distribution during the erection process of the bridge, using the free cantilevering technique for the mid-span part.

load-carrying capacity of the cantilevers. As the length of the cantilevers increased, the moment and shear over the inner supports also increased – stresses that exceed what is later the case for the final bridge (Figs. 12.4 and 12.5).

When the two cantilever ends met in the middle, the final section had to be adjusted due to the temperature deformations the bridge had been subjected to during the day. The weather had been sunny and warm, which made the bridge, in addition to the elastic bending deformations, to bend down a little bit extra. The final closing section had to be shortened 15 mm at the top (due to this additional temperature deformation) in order to fit into the gap between the cantilever ends (Fig. 12.6).

Towards the evening, when the temperature dropped, the temperature deformations reversed, however, as the final section already had been installed, a constraint was introduced. This constraint induced a tensile force in the top flange (that was prevented from shortening), and, as a consequence, compressive stresses was introduced in the lower flange. One could also say that the cantilever ends were prevented from rotating back to their original position. In addition to this effect, there were already high

Fig. 12.5 Only a short gap remains before the final closing of the two cantilevers.

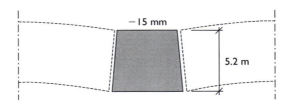

Fig. 12.6 The final closing section in the centre part had to be adjusted due to temperature deformations in the bridge.

compressive stresses introduced in the lower flange (that we also will concentrate on in the following) because of the chosen erection method. Two gigantic cantilevers had been joined together; however, the moment distribution was not yet the one that could be expected from a continuous system. And in order to achieve this, the inner supports had to be adjusted down, so that the undesired moment distribution was levelled out (Fig. 12.7).

In the late afternoon, when the two cantilever ends had been joined, there was, however, not enough time left to lower the inner supports, instead it was decided to wait until the next working day. Due to the temperature deformations described above, there was a situation in the evening that had given an *increase* in the temporary moment distribution, instead of the desired decrease that would be the case if the supports had been lowered (Fig. 12.8).

By the unfortunate decision of not immediately lowering the inner supports, and the unlucky effect from the temperature drop, there was a situation – the night of 6 November – where the lower flange was subjected to *compression* over the entire length of the bridge (due to the negative bending moment). This fact, together with

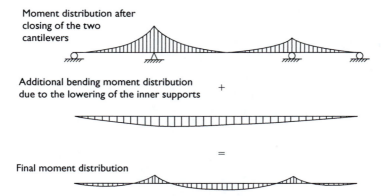

Moment distribution after closing of the two cantilevers

Additional bending moment distribution due to the lowering of the inner supports   +

=

Final moment distribution

Fig. 12.7 The moment distribution that would be the result following the *intended* lowering of the two inner supports.

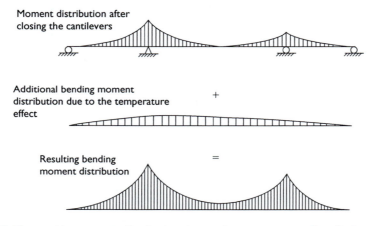

Moment distribution after closing the cantilevers

Additional bending moment distribution due to the temperature effect   +

Resulting bending moment distribution   =

Fig. 12.8 The *resulting* moment distribution due to the temperature effect (*before* the intended lowering of the inner supports).

the less good decision of choosing flat bar stiffeners, was the reason for the buckling of the bridge. A flat bar stiffener is very easily deformed during welding (due to the welding deformations coming from the excessive heat input), and can also be deformed due to more or less unavoidable hits and damages during assembly. A deformed flat bar stiffener has a markedly reduced global buckling strength capacity (see further the discussion of the Zeulenroda Bridge, Chapter 17). As the lower flange of the Danube Bridge was subjected to excessively high compressive stresses, especially in the zones where the final bending moment after lowering of the supports would have been low (see Fig. 12.7 final moment distribution) – i.e. near the zero-moment positions – it was then not surprising that buckling did occur there (the buckle farthest to the right in figure 12.9). The lower flange stiffeners at these zero-moment positions are not designed for any larger stress levels (at least not for normal in-service loadings). The other two

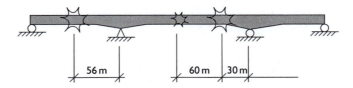

Fig. 12.9 Buckling occurred in three locations due to excessive bending moments.

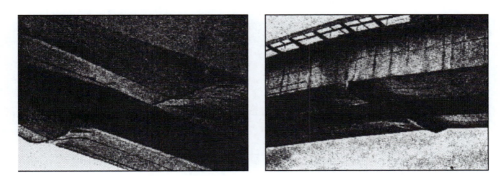

Fig. 12.10 The parts having the major buckles – close to the inner support of the main span, and in the centre part of the first span. (Aurell/Englöv: Olyckor vid montering av stora lådbroar. Väg- och Vattenbyggaren)

buckles occurred in zones where the final (desired) moment distribution would be positive, i.e. in areas where the lower flange would be subjected to tension stresses (read: in mid-span regions). However, as the moment distribution in these areas was reversed the lower flange was subjected to compression (instead of tension), and thus the stiffeners buckled due to excessive (compressive) loading (Fig. 12.9).

The buckle in the central region of the inner span came where the final section was located, however, this piece was not severely damaged (even though this section was not as strong as the rest of the bridge, due to the fact that it was assembled without the lower and upper flange plates, i.e. having only four vertical web plates). The other two buckles – the one in the main span close to the right-hand inner support (the farthest to the right), and the one in the centre part of the first span – were, however, so large that more or less true hinges were formed in these positions (inwards going buckles in the vertical web plates, and outwards going – V-shaped – buckles in the lower flanges). In the detail photos one could see the proportion of the damages in these two locations (Fig. 12.10).

In the photo of the bridge below, the deformed and upwards bent outer span can be seen to the left, as well as the downwards deflected main span (compare this deformation to the mechanism in Fig. 12.14) (Fig. 12.11).

As the continuous girder system was two times statically indeterminate, the bridge was able to withstand the formation of the two hinges. The girder system had now become *statically determinate*, with the right-hand part (1) holding up the centre part (2), which in its turn is holding up the left-hand part (3) (Fig. 12.12).

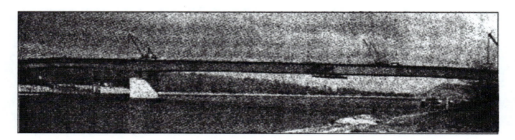

Fig. 12.11 The downwards deflected main span and the upward bent outer span (to the left). (Aurell/Englöv: Olyckor vid montering av stora lådbroar. Väg- och Vattenbyggaren)

Fig. 12.12 The statically determinate system after buckling had occurred (producing hinges in two locations).

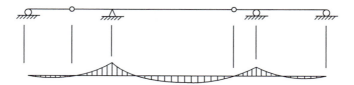

Fig. 12.13 Bending moment distribution after buckling had occurred (in fact in principle similar to the intended after the planned lowering of the two inner supports).

Fig. 12.14 The mechanism (read: collapse) of the bridge *if* the closing section also had been severely damaged.

Through this release of the inner constraint (i.e. the transformation from a statically indeterminate system to a statically determinate), the continuous girder system now received a moment distribution that was the originally intended after lowering of the supports (in principle) (Fig. 12.13).

However, if in addition the closing section also had been damaged to the same degree as the other two buckled parts – so that a third hinge had been produced – then a *collapse* (read: a mechanism) had been a fact (Fig. 12.14).

Thanks to the inbuilt damage tolerance of the system (read: robustness) the bridge could be saved, even though it was heavily distorted, and was later taken into use again (after replacing the damaged parts). The rebuilt bridge, which today goes under the name of the Prater Bridge, obtained a permanent settlement of about 700 mm in mid-span of the centre span (read: a reduced camber) as a direct consequence of the damages.

Chapter 13

Britannia Bridge

In 1850 the railway bridge over the Menai Straits in Wales was completed. The bridge was called the Britannia Bridge, and had the longest span in the world for railway traffic. The designer was Robert Stephenson, son of George Stephenson, the famous railway engineer (inventor of the record breaking locomotive "The Rocket"). It was the father who in 1838 had suggested that the railway line between London and Chester should be extended into Wales to the port of Holyhead on the island of Anglesey (to accommodate for the traffic by ferry between England/Wales and Ireland), and so it was decided (Fig. 13.1).

Instead of briefly describing the bridge – before coming to the fire in 1970 that destroyed the bridge beyond repair – we will pay homage to Robert Stephenson, and to the era when new material and new structural concepts were being tried and tested, by first giving a more detailed account of the design process of the Britannia Bridge. At the time of the destruction in 1970 a new generation of modern box-girder bridges were being built, using high-strength steel and slender plates. It took 120 years before the Britannia Bridge went down (nearly), but many of these new bridges did not even reach completion (as we shall see in Chapter 14–17, and also have learnt in Chapter 12).

The Britannia Bridge was a continuous box-girder bridge (having two separate tubes) with a total length of 460 m over four spans. The two end spans were 78 m long, and the two central spans were 152 m (often is the length of these main spans said to be 140 m, but that is only the free distance between the towers, not the supported length) (Fig. 13.2).

The somewhat excessive height of the towers was due to the original intention of having the tubes suspended with chains, but this idea was abandoned after some tests (which will be described later in this text) that showed that the tubes were strong enough to carry all load by themselves. (In a suspension bridge having a stiffening girder strong enough to take all loads, the cable becomes superfluous. A flexible cable – that changes form during loading – can only contribute to the load-carrying capacity if the stiffening girder is also flexible. Stephenson was quite rightly so aware of – after the tests – that this was not the fact in his suggestion for the bridge.)

The two end spans were assembled supported by falsework, but the two main span tubes were assembled on shore close to the bridge site (Fig. 13.3).

To erect the tubes for the main spans, the tidal water was used in an ingenious manner; the tubes were lifted from their position on shore by the help of pontoons, and then shipped out to the bridge location. By the help of hydraulic presses the tubes were then lifted up to the towers – which had been prepared with temporary channels

Fig. 13.1 Great Britain and Ireland. Menai Straits between the Island of Anglesey and the mainland of Wales is encircled.

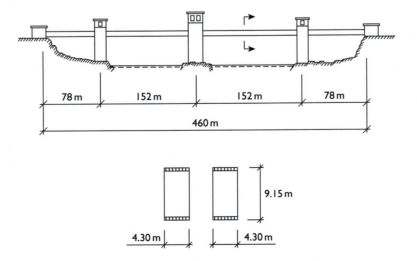

Fig. 13.2 The Britannia Bridge – elevation and cross-section.

Fig. 13.3 The two main span tubes were assembled on shore. (Ryall: Britannia Bridge, North Wales: Concept, Analysis, Design and Construction. International Historic Bridge Conference 1992)

Fig. 13.4 Preparation for erection of one of the main span tubes. (Hopkins: A Span of Bridges – an illustrated history)

in the exterior face to accommodate for the tubes (which, of course, were longer than the free spacing between the towers) (Fig. 13.4).

The separate tubes were then joined together by lifting one end of a tube up, and then connecting the opposite end with the adjacent tube (see Fig. 13.5). In this way continuity in the system was achieved, which had never been done before for a multi-span girder bridge. By lifting the ends before the joining, and by doing so prestressing the structure, continuity was achieved also for the self-weight, not only for the traffic load (which had been the case if the ends would have been joined without the lifting procedure). Stephenson had designed the box-girders as simply supported, but knew

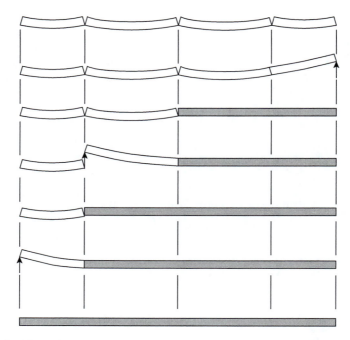

Fig. 13.5 By lifting the ends of the tubes before joining them together the tubes became prestressed, i.e. acting as a continuous system.

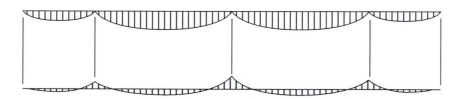

Fig. 13.6 The difference in bending moment distribution before and after prestressing.

that extra safety and load-carrying capacity would be achieved through this measure, not to forget also a reduced deflection.

The moment distribution for self-weight (and additional traffic load) with and or without prestressing shows the great difference in behaviour and stress level (i.e. between a simply supported system and a continuous) (Fig. 13.6).

The cross-section of the box-girders is stiffened in top and bottom with a number of closed cells (small box sections, parallel to the longitudinal axis, integrated with the large cross-section), eight in the top flange and six in the bottom flange. These cells, which are large enough to accommodate for the re-painting inside, are made out of plane plates, angle irons (i.e. L-profiles) and cover plates, which all have been riveted together in pre-punched holes. These closed cells act as efficient stiffeners with respect to normal stress buckling (due to the bending moment the girders are subjected

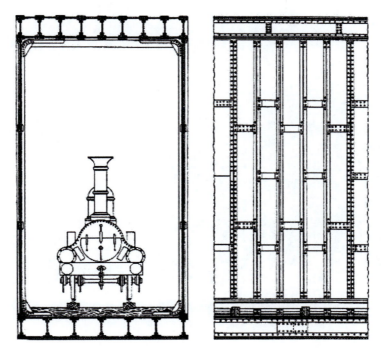

Fig. 13.7 Cross-section and side view of the tube wall. (Werner: Die Britannia- und Conway-Röhren-brücke).

to) – both for the cross-section as a whole, and also for the small-scale cells (i.e. local buckling of the plane plates in between the angles). One could also see the cells as flanges for the box-girder cross-section, consisting of double plates, vertically stiffened inside. In the bottom flange the cells also have to carry the transverse (vertical) loading from the railway traffic (Figs. 13.7 and 13.8).

The tube walls are made up of plane, rolled plates of wrought iron (having a thickness of 16 mm or smaller), which have been spliced together, in the vertical direction by T-profiles, and in the longitudinal direction by cover plates (which had to be bent at the edges in order to wrap over the flanges of the T-profiles). The longitudinal splices are shifted in position in between adjacent plates. In the figure below the individual parts are separated from each other in order to show the built-up system (Fig. 13.9).

As the vertical splice is also stiffened inside with a T-profile, the resulting vertical stiffener is made double-sided, and thus very strong (Fig. 13.10).

Even if the compact flanges take the major part of the bending moment, the walls do contribute to some extent, which make them also subjected to normal stresses that can be a local buckling risk. The closely spaced T-profiles (610 mm), together with the chosen wall plate thickness, are, however, a sufficient barrier against normal stress buckling, even if horizontal stiffeners are the normal choice today. In a box-girder bridge, having a cross-section depth of over nine meters, the expected wavelength of a horizontal buckle (in the post-critical range) in the compression zone would be several

Fig. 13.8 Elevation of the Britannia Bridge. (http://vivovoco.rsl.ru/VV/E_LESSON/BRIDGES/
BRIT/BRIT.HTM)

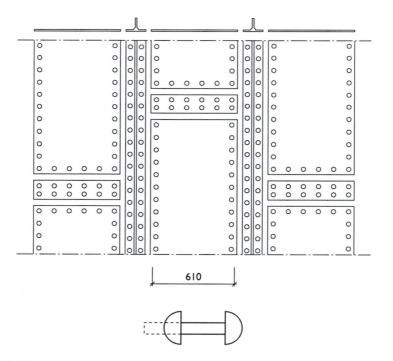

Fig. 13.9 The individual plates and T-profiles of the tube wall before being riveted together.

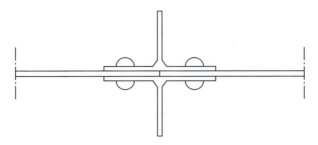

Fig. 13.10 The vertical splice T-profiles provided a strong vertical stiffening of the wall.

meters long (approximately 2/3 of the depth, given that the web is unstiffened). This fact does show that the close spacing between the vertical stiffeners of the Britannia Bridge is more than sufficient. Vertical buckling of the web plate due to concentrated loads is also not a problem, as the load from the trains is introduced in the bottom part of the box, and thus creates only vertical tension as the load is transferred into the web plate. With respect to shear buckling of the walls, the chosen vertical stiffeners are in good agreement with the principles regarding how this type of buckling phenomenon is handled today – crossing diagonals, in tension and compression, in each panel in between vertical stiffeners carries the shear stress flow. The wavelength here is even longer than for normal stress buckling – approximately 1.25xh – which makes the closely spaced vertical stiffeners also act as a sufficient barrier against shear buckling. However, the choice of vertical stiffeners in the Britannia Bridge was criticized by the Russian bridge and railway engineer D. J. Jourawski. He claimed that if the stiffeners had been positioned in the compression diagonal, having an inclination of 45° towards the longitudinal axis, the efficiency would be much greater than for the vertical stiffener configuration that was chosen by Stephenson.

When the trains were to pass the bridge, the tubes could obviously not be stiffened inside by cross-framing, instead the torsional stiffness had to be achieved by a horizontal stiffening truss in the roof, and in addition, all four corners of the box were stiffened in order to ensure frame action.

It was the tough requirements from the Admiralty regarding free sailing space in the horizontal and vertical direction that made the choice of a (suspended) girder bridge inevitable (suspended at least according to the original concept). The demands were 32 m free height and 137 m free width in the two sailing channels underneath the bridge. In addition the channel was to be kept open for the passage of ships during the construction. These demands taken together made it impossible to consider the otherwise obvious solution, which would have been a cast iron arch bridge. Besides thinking in terms of a totally new concept for the bridge, they had also to abandon the thought of cast iron as a possible construction material. Cast iron is as hard as (mild) steel, but much more brittle (i.e. less ductile), which makes it suitable for arches where the compression forces are dominating, but less suitable for structures where tension is also present, which definitely would be the case here. Wrought iron, however, is as tough in both compression as in tension, and not as brittle as cast iron. The reason why wrought iron was not as common and competitive as a construction material for bridges was the much higher cost and the fact that it was more difficult to prepare the plates into desired shapes. Cast iron profiles are made directly from the moulding form, which is a huge advantage, however, thus receiving a brittle material. Stephenson pondered on the solution of having two parallel I-girders standing next to each other – clearly inspired by the simple I-girder bridge concept that was so common at the time for short and medium span bridges – but having the top and bottom flange here connected to each other. The shorter I-girder bridge spans were without exception in cast iron, and had a maximum span length of 15–20 m. The medium span bridges had extra stiffening bars of wrought iron on the tension side, and were approximately 30 meters long. For longer spans the girder bridge had to be suspended with wrought iron eye-bar chains. Stephenson was fully aware of the flexibility of these latter structures (for wind and heavy loading, and not to forget about marching troops!), and knew that his structure had to be a (suspended) stiff girder bridge.

The strength and stiffness of deep plates was confirmed to Stephenson as he was reported about an incident when the steam vessel Prince of Wales had been launched. The ship had accidentally become hanging with the bow in the water and the stern still supported by the slipway on shore, which made the ship's hull becoming simply supported over a length of more than 30 meters, however, without giving any major damages to the hull.

Together with the ship-builder W. Fairbairn and the mathematician E. Hodgkinson, Stephenson carried out a number of experiments in order to determine the action and load-carrying capacity, not only of rectangular cross-section girders, but also of circular and elliptical girder tubes. It was mainly the strong concern he had regarding the wind loading that made him consider the two latter choices (which were more aerodynamically shaped). As the first thought of a possible cross-section shape went to these last mentioned profiles, it is easy to understand where Stephenson did get the idea for hollow tubes, as the trains had to pass inside somehow. Girders in a model scale were tested in three-point bending (i.e. simply supported with a point load in mid-span). They found very soon that the rectangular shape was superior to the other two, mainly due to the buckling tendency of the latter. The flange in a rectangular cross-section is both larger and more efficient than for a circular or an elliptical cross-section. It was also in these tests that they found that the flange in compression had to be shaped with closed cells in order to minimize the risk of buckling, and by doing so ensuring a load-carrying capacity as close to the base material as possible. Fairbairn, who earned a reputation as being methodical and meticulous concerning the different research investigations he did undertake, did comment:

> "Some curious and interesting phenomena presented themselves in the experiments – many of them are anomalous to our preconceived notions of the strength of materials, and totally different to anything yet exhibited in any previous research. It has invariably been observed, that in almost every experiment the tubes gave evidence of weakness in their powers of resistance on the top side, to the forces tending to crush them."

Fairbairn was quite clearly referring to earlier experiments on cast iron girders, which had without exception gone to failure on the tension side (as a brittle fracture), while these tests on wrought iron girders had the failure coming on the compression side (as normal stress buckling). Hodgkinson made the same observation:

> "It appeared evident to me, however, that any conclusion deduced from received principles, with respect to the strength of thin tubes, could only be approximations; for these tubes usually give way by the top or compressed side becoming wrinkled, and unable to offer resistance, long before the parts subjected to tension are strained to the utmost they would bear."

Hodgkinson knew that the simple mathematical expressions which traditionally had been used – where the maximum load-carrying capacity was derived from the ultimate *tensile* strength of the material – definitely overestimated the capacity. It was also during this work, together with additional buckling tests they made on plane plates, that Hodgkinson found that the maximum load-carrying capacity, with respect on buckling of an axially loaded plate, was directly proportional to the thickness of the plate in cube. That this is the case is easily derived if we transform the expression for the critical buckling stress into an equivalent axial load (Eq. 13.1):

$$\sigma_{cr} = k \cdot \frac{\pi^2 \cdot E}{12 \cdot (1 - \upsilon^2) \cdot \left(\dfrac{b}{t}\right)^2}$$

$$\Rightarrow \quad P_{cr} \sim t^3 \qquad\qquad (13.1)$$

$$P_{cr} = \sigma_{cr} \cdot b \cdot t$$

Hodgkinson had clearly an early and genuine understanding of the ultimate capacity of axially loaded plates, even though that he stated that the capacity was limited by buckling (and therefore excluded the post-critical reserve strength). And by excluding the reserve effects in the design of the Britannia Bridge, the structure came to have an extra safety level that to some extent explains why it was possible to raise the traffic load without having to strengthen the bridge.

Stephenson and Fairbairn also found in the experiments that the clamping force in between the connected plates was sufficiently high, that is that the friction was not exceeded due to shearing in the joints. The deflection one could expect from the bending deformations of the box-girder spans was consequently only coming from elastic deformations of the material, and not by additional movements from shearing making the rivets coming in contact with the rivet hole edges. This fact was also confirmed as the deflection was checked during the passage of trains. Despite small expected deflections one still chose to provide the centre spans with a camber of 23 centimetres.

The concept of having a closed box-girder section, where especially the walls were made out of solid plates, made the bridge become extremely expensive due to the large amount of material that was used. The labour cost during the 1800's was low, but the cost of the material was high. The development instead went therefore towards more lighter and material-saving truss structures – a development that started in the USA. Smaller truss bridge spans in wood became increasingly longer as more and more parts were replaced by iron. A special concept that was invented by Pratt, where the diagonals were positioned so that they always became in tension during loading,

Fig. 13.11 The box-girder bridge at Conway, 25 kilometers east of the Menai Straits.

was very competitive, as material was saved due to the non-existent risk of global buckling of these diagonals. Stephenson, however, carried on by designing and developing his heavy and robust box-girder bridge concept with great success during the 1850's. Among others, he designed for example the Victoria Bridge (1859) over the St. Lawrence River near Montreal in Canada (see also Chapter 4). This bridge was the longest bridge in the world at that time (1.8 kilometers long), and did have a maximum span length of 100 meters. The bridge had to be rebuilt after some time though, as the smoke from the locomotive troubled the passengers – the walls had to be opened up!

Even though that the Britannia Bridge is said to be the start of the era concerning box-girder bridges, the fact is that a box-girder bridge was built in Conway – 25 kilometers east of Menai Straits, also on the Chester-Holyhead railway line – already a year earlier. This bridge, with a single span of quite impressive 122 meter, served as a prototype for the Britannia Bridge – here Stephenson had the opportunity to test the design and erection method. It was from the Conway Bridge that the idea came for the camber of the Britannia Bridge (the Bridge at Conway was made completely straight, which made the deflections from self-weight and traffic become visible). It was also some important experiences made from the lifting process, even though the lifting height at the Conway Bridge was only 5.5 meter, and not 36 meter as was the case for the Britannia Bridge (Fig. 13.11).

The Conway Bridge was reinforced in 1899 with two extra inner supports, approximately 15 meters from each abutment, which made this bridge also continuous. The extra supports can be seen in the photo above, in between the castle tower supports.

The Britannia Bridge was the longest box-girder bridge in the world (given the maximum span) up until the period after the Second World War, when several new and

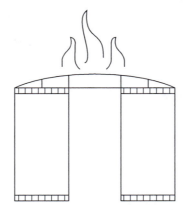

Fig. 13.12 The wooden roof, which protected the two tubes, was put on fire.

Fig. 13.13 The burnt-down roof of the Britannia Bridge. (www.2d53.co.uk/britanniabridge/menu.htm)

modern box-girder bridges were built, mainly in post-war Germany. The Britannia Bridge carried heavy and modern traffic loading without having to be strengthened, and continued to do so until 1970 (after 120 years of faithful service), when the wooden roof – which had been added to the original structure to protect against rain – accidentally was put on fire. On the evening of 23 May 1970, two boys, looking for bats close to the entrance at the south end of the bridge (at the mainland), dropped a burning torch which first ignited the tar-coated roofing felt, and then spread to the timber boards (Fig. 13.12).

All through the night the fire brigade tried in vain to extinguish the fire, but eventually they had to withdraw, accepting that the fire spread over the entire length of the bridge towards the Anglesey side (the north end). On the next day all that was remaining of the roof was the steel L-profiles (Fig. 13.13).

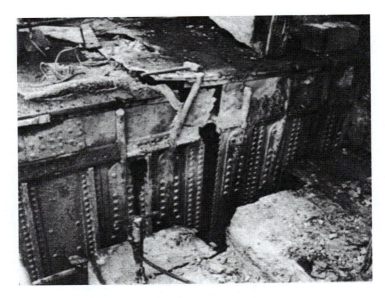

Fig. 13.14 The tubes split open over each intermediate support (i.e. at the towers).

The bridge was still standing, but the tubes had cracked open over all three intermediate supports, i.e. in the maximum bending moment regions, and the spans were visibly sagging (Figs. 13.14 and 13.15).

The two end spans had smaller deflections, but the two main spans were deflecting 49 and 71 centimetres respectively. The statically indeterminate system had been transformed into a statically *determinate* system (i.e. becoming four simply supported box girders, instead of the continuous system it originally was). Perhaps it was the decision by Stephenson in 1850 to prestress the tubes that saved the bridge from a total collapse (see Fig. 13.5). Possibly the sagging did not consist only of the spans being simply supported, but most probably irreversible plastic deformations due to the heating of the tubes contributed as well. As the upper part of the tubes (i.e. the top flange) was heated, the wrought iron material softened, resulting in a decrease in stiffness, which reduced the effective cross-section of the tubes as well as lowering the neutral axis. The stresses on the upper tension side (for the negative bending moment regions over the intermediate supports) thus increased markedly (Fig. 13.16).

As a result of this combined effect there is a high probability that the tension stresses exceeded the yield limit of the material, leaving irreversible plastic strain deformations in these parts. If the tension stresses were even higher it could also explain why these regions cracked open, but most certainly the cracking was due to restricted contraction during cooling (almost the same mechanism as why micro-cracks in welds are being formed). When the tubes became hot, and the deflections increased, this resulted in large tensile strains in the upper half of the tubes (over the supports). As the fire died out, and the wrought iron material wanted to contract, it was then restricted by

Fig. 13.15 The sagging of the bridge spans because of the vertical split over the intermediate supports. (http://vivovoco.rsl.ru/VV/E_LESSON/BRIDGES/ BRIT/BRIT.HTM)

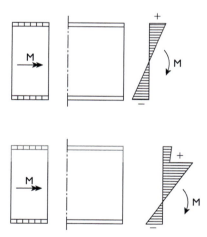

Fig. 13.16 As the upper part was heated, and lost its stiffness, the tension stresses for the negative bending moment regions increased – both because of the reduced net-section and also because of the neutral axis being lowered.

cooler parts and from the weight of the deflected spans. When the stiffness gradually increased (due to the cooling) the elongation that was forced upon the upper part resulted in fracture stresses. As a similitude we could consider a metal bar being put in an oven – the bar expands as the temperature increases, but will contract again as

Fig. 13.17 Aerial view of the Britannia Bridge of today. In the background the Menai Strait Suspension Bridge built by Thomas Telford in 1826 can be seen. (http://vivovoco.rsl.ru/VV/E_LESSON/BRIDGES/ BRIT/BRIT.HTM)

soon as the temperature is being lowered, without any stresses being formed. However, if the contraction for some reason is restricted (due to some constraints holding it back during cooling) then tensile stresses would be built into the bar (even leading to fracture if the tensile strain becomes too high). There is, however, also the possibility that the wrought iron material cracked already during the attempts of the fire brigade to extinguish the fire – applying cold water on hot iron.

The permanent deformations and the apparent risk that the tubes would loose their supports over the piers, made the authorities decide that the Britannia Bridge had to be rebuilt. Many different solutions to save the bridge were considered but, finally, it was decided upon to reconstruct the entire bridge completely, where the two main spans were being rebuilt as a truss arch bridge in steel. The choice of using truss arches was two-folded; first it was an idea that already back in the 1840's had been proposed by Stephenson, but rejected by the Admiralty because of the free sailing width requirements, and secondly it enabled for the rapid use of one of the two parallel tubes (being supported from beneath) for traffic, while the other tube (also that supported) was dismantled. The bridge was finally opened for regular traffic in 1974, and although being completely reshaped (using steel arches to carry a new concrete deck to support the railway track, the walls and roof of the existing bridge taken away) the original towers and abutments were used with only minor adjustments. The two end spans were constructed as steel girder bridges supported by two concrete frames each. To accommodate for highway traffic as well, a second deck (supported by steel frames) was constructed and put on top of the railway deck in 1980. The nearby Menai Strait Suspension Bridge from 1826 had over a long period of time created traffic congestions, and it was therefore decided to transform the Britannia Bridge into a two-level bridge. So today the Britannia Bridge – in its new shape – is carrying railway traffic on the lower level, as it always have done, and highway traffic on the upper level (what earlier was the roof of the original structure) (Fig. 13.17).

It was definitely not a case of a total demolition of a unique engineering monument – it was rebuilt because of the damages of the fire and due to the need to meet modern standards, and, not to forget, re-using one of the original ideas proposed by Stephenson. As a train passenger you are also travelling at the same height and position as before, knowing that the Britannia Bridge have, in this location, been carrying railway traffic since 1850, and, as a bonus, now enjoying a view that earlier was restricted because of the solid plate walls. And finally, not to forget, the tubular bridge at Conway is still standing intact in its original design (similar to the old Britannia Bridge)!

Chapter 14

Cleddau Bridge

Less than seven months after the buckling incident of the Fourth Danube Bridge (see Chapter 12) the Cleddau Bridge, near the seaport of Milford Haven in Wales (Fig. 14.1), collapses on 2 June 1970 during construction. It is almost by irony of fate that just ten days after the destruction of the Britannia Bridge (see Chapter 13) the Cleddau Bridge collapses – a bridge that would be one of the longest in Europe (having a maximum span of 214 metres). The predecessor and the inheritor are made unfit for use almost at the same time, and top of all, they were both located in Wales.

The Cleddau Bridge was to be a continuous box-girder bridge in seven spans, with a total bridge length of 820 metres (in the redesigned bridge though – after the collapse – the centre part was rebuilt as a cantilever bridge, having reduced span lengths) (Fig. 14.2).

After the end spans on either side had been erected (using falsework supports), the free cantilevering erection method was used for the continued assembly. For the north end (to the left in the elevation) it was decided to use a temporary support in mid-span of the second span, while on the south side (to the right in the elevation) it was decided to manage without, due to the shorter span length. The trapezoidal cross-section of the bridge had a lower flange width of 6.7 metres, an upper flange width of 20.1 metres, and a depth of 6.1 metres (Fig. 14.3).

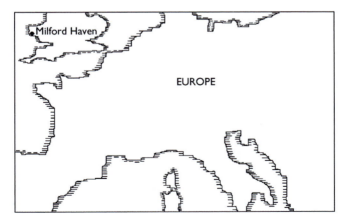

Fig. 14.1 The Cleddau Bridge was located in the southwestern part of Wales in Great Britain (see also Fig. 13.1).

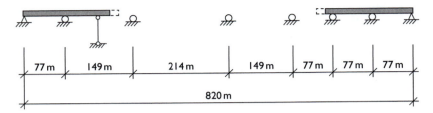

Fig. 14.2 Elevation of the Cleddau Bridge during erection (prior to the collapse of the cantilever arm at the south end of the bridge – on the right-hand side in the figure).

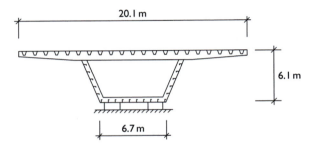

Fig. 14.3 Cross-section of the Cleddau Bridge.

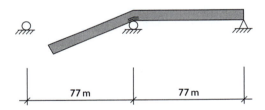

Fig. 14.4 Collapse of the second span on the south end of the bridge.

The designer, Freeman, Fox & Partners, had with great success a couple of years earlier used a trapezoidal cross-section at the construction of the Severn Bridge (a 988 metre long suspension bridge) and at the construction of the Wye Bridge (a 235 metre long cable-stayed bridge), and wanted to continue with this successful concept. The box-girder bridge cross-section of the Cleddau Bridge was – in contrast to the Fourth Danube Bridge – stiffened in the longitudinal direction by the use of more buckling stiff and torsion-rigid stiffeners. The assembly of the second span on the south end of the bridge, was just before the collapse cantilevering out 61 metres, and the final section (closing the gap of 77 metres) was just to be launched out as the bridge buckled over the inner support, and the huge arm fell 30 metres to the ground (killing four people) (Figs. 14.4 and 14.5).

In contrast to the Danube Bridge – which luckily received its damages first *after* the system had become statically indeterminate (i.e. being closed at midspan) – the Cleddau

Fig. 14.5 The cantilever arm of the second span on the south end of the bridge buckled over the inner support and fell to the ground. (Aurell/Englöv: Olyckor vid montering av stora lådbroar. Väg- och Vattenbyggaren)

Bridge had no such extra inbuilt safety during construction, as the system still was statically *determinate*. As soon as this hinge was produced (read: the buckling at the inner support) a mechanism became the result. At the investigation it was established that an inadequately stiffened diaphragm had initiated the buckling over the support.

Diaphragms are made up of a solid plate in box-girder bridges (having for a centrally located manhole though, for the passage of inspection personnel), and these are positioned at a regular interval in the longitudinal direction (normally 8–12 m) in order to ensure a sufficient torsional stiffness of the cross-section. The diaphragms are stiffened, especially over the supports, as they are in these positions subjected to large concentrated forces. In the Cleddau Bridge, the diaphragm plate was stiffened with so-called bulb-flat stiffeners, having the dimension $250 \times 13 \, \text{mm}^2$. The plate itself was 10 mm thick in between the bearings, and 13 mm locally over the bearings and for the inclined outer parts (Fig. 14.6).

A diaphragm over a support has many functions:

– Transfer the shear load, that is coming from the inclined webs, down to the bearings,
– Carry the reaction forces,
– Act as a vertical stiffener of the web plate,
– Act as a stiffening girder when a bearing has to be replaced,
– Transfer the horizontal forces from wind and traffic down to the bearings.

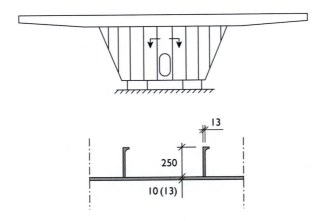

Fig. 14.6 The stiffened diaphragm plate of the box-girder over the inner support.

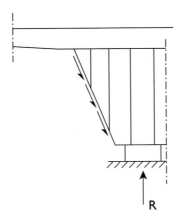

Fig. 14.7 The shear load from the inclined webs is transferred down to the bearings.

The first function mentioned above, means that the diaphragm has to act as a stiffening (transverse) girder – supported by the two bearings – which is loaded at the edges by the inclined shear force (Fig. 14.7).

As a comparison – in order to better understand the action of a diaphragm having inclined edges – we could study a diaphragm having a rectangular cross-section (Fig. 14.8).

The shear load must also in this case be transferred from the edges (where the web plates are connected) to the bearings. In a simplified manner we assume that all loads are introduced in the upper part of the diaphragm, and are then transferred down to the bearings through a diagonal in compression. And even if diaphragms in general are large, there are locally concentrated forces in compression over the bearings. Besides the vertical reaction force and the diagonal force, the inclined diagonal force has also to be balanced by a horizontal force in this region (Fig. 14.9).

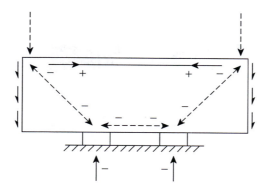

Fig. 14.8 As a simplified model showing the load distribution – a rectangular diaphragm plate.

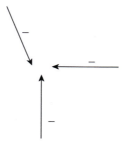

Fig. 14.9 Equilibrium due to the inclined diagonal force is met by a horizontal balancing (compressive) force.

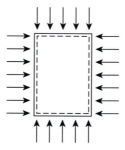

Fig. 14.10 Locally, close to the bearings, the diaphragm plate is subjected to two-axial compression.

The diaphragm plate is thus subjected to a two-axial compression locally over the bearings (Fig. 14.10).

The critical buckling stress for a plate loaded in two-axial compression, follows a linear relationship according to (Eq. 14.1):

$$\frac{\sigma_x}{\sigma_{x,cr}} + \frac{\sigma_y}{\sigma_{y,cr}} = 1.0 \quad \Rightarrow \quad \sigma_x = \sigma_{x,cr} \cdot \left(1 - \frac{\sigma_y}{\sigma_{y,cr}}\right) \qquad (14.1)$$

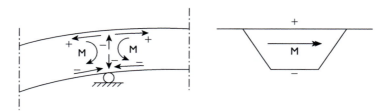

Fig. 14.11 Additional vertical reaction force (in the diaphragm plate) was induced due to the curvature effect.

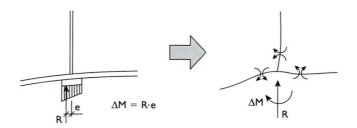

Fig. 14.12 Eccentric loading of the bearing pad induced local bending in the lower flange plate and in the diaphragm plate.

The relationship could also be expressed through the buckling coefficient k (Eq. 14.2):

$$k_x = 4.0 \cdot \left(1 - \frac{\sigma_y}{\sigma_x}\right) \tag{14.2}$$

As the loading in the transverse direction is increasing, the critical buckling stress – for the higher loading in the x-direction – will decrease (i.e. the buckling coefficient will become less than 4.0).

Besides the two-axial stress state – due to the transfer of the shear force from the inclined webs – there were a number of other factors that even more increased the stresses over the support. There was, for example, the additional compression in the diaphragm plate caused by the curvature (coming from the bending deformations because of the deflection of the cantilever arm) (Fig. 14.11).

In addition, the 40 mm wide bearing plate also became eccentrically loaded as a local eccentricity was induced because of the curvature. This eccentricity made the lower flange plate (of the box-girder cross-section) and the diaphragm wall to become subjected to a bending rotation (Fig. 14.12).

In the investigation it was also found that the "holding-down bolts" – which at a later stage of the erection (cantilevering of the third span) should prevent lifting of the bearings – had been tightened. The effect of this tightening probably made the supports becoming fixed, which additionally increased the vertical reaction force over the supports (Fig. 14.13).

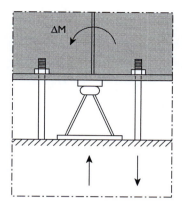

Fig. 14.13 Tightening the bolts (which were needed in holding down the box-girder bridge first when cantilevering out the *third* span) already at this early stage meant that an additional reaction force was introduced (due to the clamped condition).

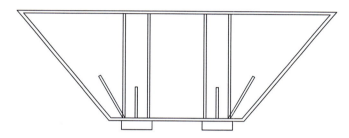

Fig. 14.14 The configuration of a *modern* box-girder bridge diaphragm plate, having extra stiffeners in the highly stressed regions.

All these effects taken together created excessive loading of the diaphragm locally over the supports, which made the diaphragm plate and its stiffeners buckle, followed by the buckling of the web plates, and ultimately the collapse of the entire cross-section. The design codes at that time was quite clearly inadequate with respect to proper construction of diaphragm plates, and the designers were also not fully aware of the complexity regarding the action of the same. Today the diaphragms are not only having larger thicknesses and stronger stiffeners, they are also locally provided with extra stiffeners where the stress concentrations are the highest (Fig. 14.14).

Chapter 15

West Gate Bridge

The West Gate Bridge in Melbourne (Fig. 15.1) was to become without comparison the largest bridge in Australia – four lanes in each direction, and with a total length of almost 2.6 kilometres.

The bridge consisted of two parts; approach spans in reinforced concrete of 67 metres each, and a centre part in steel having a total length of 848 metres. This steel section was a continuous box-girder bridge in five spans, having the three centre spans – over the Lower Yarra River – suspended by stay-cables. The main span of 336 metres would be one of the longest in the world for a cable-stayed bridge (Fig. 15.2).

The trapezoidal cross-section had four webs – two inclined and two vertical – all together forming a three-cell box-girder bridge. Besides contributing to the load-carrying capacity, the vertical webs also functioned as inner supports for the crossbeams in the upper and lower flange (Fig. 15.3).

The choice of designer once again fell on the renowned English consulting engineer Freeman, Fox & Partners, as the competence was judged to be lacking in Australia for such a huge bridge. FF&P also had an impeccable and good reputation dating back to 1932 when being the designers of the successful Sydney Harbour Bridge.

The construction of the bridge started in April 1968, and it was a hoped that the work should be finished in December 1970, however, due to strikes and other delays,

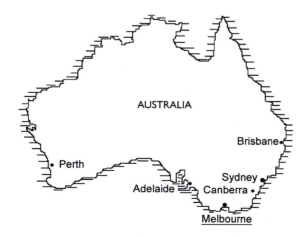

Fig. 15.1 The West Gate Bridge was located in Melbourne, in the southernmost part of Australia.

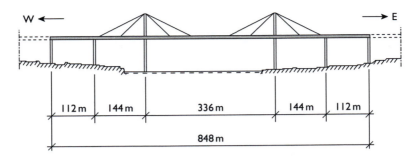

Fig. 15.2 Elevation of the West Gate Bridge (the central steel part).

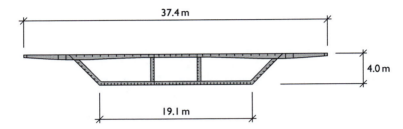

Fig. 15.3 Cross-section of the West Gate Bridge.

the work came to be seven months behind time schedule already by the end of 1969. In the beginning of 1970 the original steel contractor was replaced, but the fact still remained that construction was much overdue. The stress was definitely not lessening as the message arrived in June 1970 telling that the Cleddau Bridge in Milford Haven had collapsed, and that FF&P was the designer for that bridge as well (see Chapter 14). FF&P made the best of the situation, trying to explain to the authorities and the working personnel that it was a once in a lifetime incident, and that the bridge in Milford Haven was built using the free cantilevering technique, which was not the erection method used for the West Gate Bridge. The first spans on the east and west sides (those having span lengths of 112 metres) were just to be erected, and the technique chosen was to build the spans on the ground before lifting them into position resting on the piers. The Milford Haven collapse, however, led to the strengthening of these cross-sections before lifting commenced.

The time pressure made the contractor choose a somewhat unusual erection sequence though. In order to save time and reduce the weight for the lifting process, it was decided to build the girder spans in two separate halves. One half at the time was then to be lifted up by the help of hydraulic jacks, and then launched horizontally on a sliding girder into position (Figs. 15.4 and 15.5).

Already at the assembly of the first half of the bridge, on the ground on the east side, problems had arisen. The free flange edge of the inner part had buckled, when the span lay simply supported before the lifting process, having the entire length free and unsupported. The box-girder half became already on the ground subjected to maximum moment due to self-weight – the lower flange in tension, and the upper flange in compression, which buckled due to the stresses it became subjected to (Fig. 15.6).

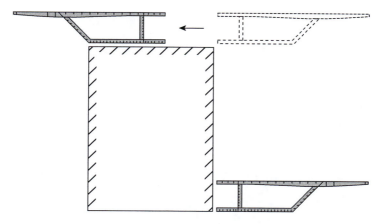

Fig. 15.4 The lifting procedure of the two separate halves.

Fig. 15.5 The two separated halves, on top of the supports, before being joined together. (Report of Royal Commission into the Failure of West Gate Bridge. Copyright State Government of Victoria, Australia)

Strangely enough did they not attend to the problem when the bridge half was still on the ground, instead it was decided to continue with the lifting process, and *thereafter* deal with the buckled flange edge up in the air. The logical thing had of course been to remove the stresses by supporting the girder span along the *entire length*, and thereby unload and straighten the upper flange. And by doing so, giving them the chance of strengthening the flange edge *before* the lifting procedure started.

The decision to assemble the bridge in two separate simply supported halves was the reason why the flange became free and unsupported, which also made it buckle. We will in the following study this free flange edge more in detail, and compare the

Fig. 15.6 The upper (and outstanding free) flange buckled already when the bridge half lay on the ground, supported at the ends (before being lifted). (Report of Royal Commission into the Failure of West Gate Bridge. Copyright State Government of Victoria, Australia)

critical buckling stress of the same in comparison to the case when the two flanges of the separate halves have been joined together. The protruding free flange was reinforced in the longitudinal direction by bulb flat stiffeners (just as for the Cleddau Bridge in Milford Haven) each 1060 millimetres. At the free edge of the flange, this distance came to be halved. The flange plate and the longitudinal stiffeners were in their turn stiffened by a crossbeam every 3.2 metres in the transverse direction – the cross-beam was also a bulb flat stiffener (460 millimetre deep) (Fig. 15.7).

Let us now study the part of the stiffened flange that is closest to the free edge, having the remaining three edges supported by a longitudinal stiffener and two crossbeams (Fig. 15.8).

The critical buckling stress (for this plate supported on three edges only), with respect to an evenly distributed compressive load, will be (Eq. 15.1):

$$\sigma_{cr} = 0.425 \cdot \frac{\pi^2 \cdot E}{12 \cdot (1 - v^2) \cdot \left(\dfrac{530}{9.5}\right)^2} \tag{15.1}$$

As the free edge will be joined with the opposite flange plate (from the other bridge half), the critical buckling stress will then *increase* (i.e. beneficial), even though the width has been doubled (Fig. 15.9 and Eq. 15.2):

$$\sigma_{cr} = 4.0 \cdot \frac{\pi^2 \cdot E}{12 \cdot (1 - v^2) \cdot \left(\dfrac{2 \cdot 530}{9.5}\right)^2} \tag{15.2}$$

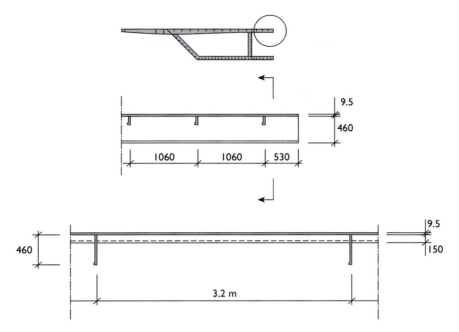

Fig. 15.7 Cross-section of the bridge half, with a close-up of the upper flange end together with a side-view of the same.

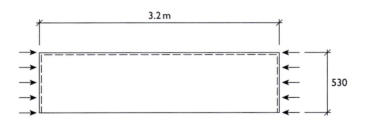

Fig. 15.8 The free flange plate in between longitudinal and transverse stiffeners.

If we compare the critical buckling stresses with each other, we find that the latter – the one with the doubled width – is 2.35 times larger than the former (Eq. 15.3):

$$\frac{\left(\dfrac{4.0}{2^2}\right)}{0.425} = 2.35 \tag{15.3}$$

The critical buckling stress became more than 50% reduced due to this temporary free edge of the flange, which also became apparent already on the ground. However, as has been mentioned earlier, they waited until the bridge was lifted into position before taking care of the buckles, even though the buckles were up to 380 millimetres in amplitude! The reason behind the large amplitudes was the fact that also the

Fig. 15.9 The width of the flange plate – in between the longitudinal stiffeners – will be doubled as the two bridge halves are joined together, but still the critical buckling stress will increase.

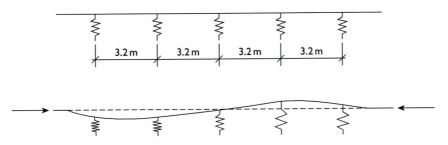

Fig. 15.10 Buckling of the longitudinal stiffener, being supported by flexible transverse cross-beams.

longitudinal stiffener (closest to the free edge) had buckled, due to the flexibility of the transverse cross-beams (which did not act as rigid supports, as they were flexible cantilevers) (Fig. 15.10).

This flexibility of the transverse cross-beams made the longitudinal stiffener have a buckling length longer than the distance between the cross-beams. The deflections of the cross-beams were 50–75 millimetre, and it was this flexibility that contributed to the buckling of the plate edge, and also explains why the deflection of the same became as large as 380 millimetre. In addition, a weak splice of the longitudinal stiffener also contributed to the reduced buckling strength. Each 16 metres the longitudinal stiffeners were spliced by using a simple single-lap joint, having a rectangular plate $100 \times 12.5\,mm^2$ overlapping the two bulb-flat stiffener edges. By this choice of configuration a local weakness was inserted – the overlapping plate was not only smaller in dimension (than the bulb flat stiffeners), it was also eccentrically positioned, and last, but not least, not welded to the upper flange plate, which made it less strong in the transverse out-of-plane direction (the same applies for the deck plate). This detail had thus a markedly reduced strength in transferring axial load from one end of a longitudinal stiffener to the other (Fig. 15.11).

Being up in the air, there was no chance of unloading the bridge half, instead they had to solve the problem with the buckled plate edge in a different manner. A bold decision was then taken, and that was quite simply to open up transverse splices in

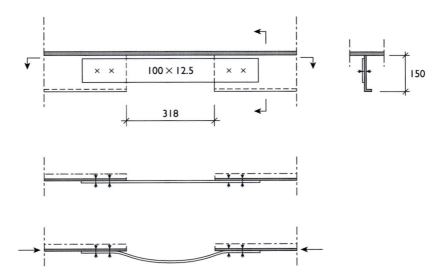

Fig. 15.11 The rather weak detailing of the longitudinal stiffener splices (each 16 metres).

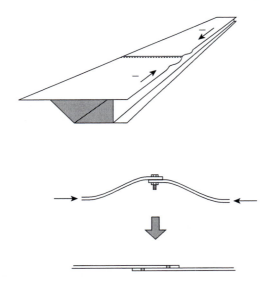

Fig. 15.12 By removing the bolts in transverse splices (of the deck plate) the buckles were removed when the plates slipped under each other.

order to "relieve" the inbuilt stresses, which were the cause of the buckling – the plates would then slip under each other, and by doing so straighten out the buckled plate. This was a step that worked quite satisfactorily for the east span (Fig. 15.12).

New holes had to be drilled – or the enlargement or re-drilling of already existing – in order for the bolts to be inserted again (note that the size of the buckle and consequently the slip between the plates is heavily exaggerated in the drawing above).

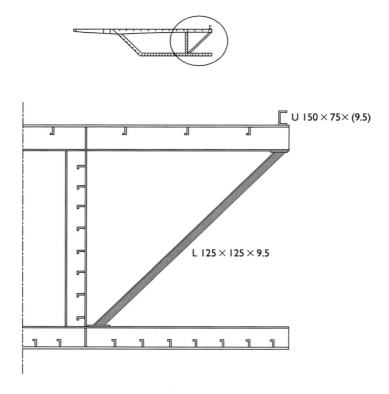

Fig. 15.13 The upper flange edges of the bridge halves on the west side were stiffened by an extra longitudinal, and the transverse cross-beams were supported by an additional diagonal.

For the continued assembly of the bridge halves on the *west* side, the free flange edges were stiffened with an extra longitudinal, and each cross-beam was supported by a diagonal – all due to the lessons made from the buckling of the east girder span (Fig. 15.13).

At the assembly of the west girder halves in the air, this extra reinforcement proved to be efficient, and there was initially no problem with buckling of the upper flange. Instead another problem arose when the two halves were to be joined together – due to tolerances that were exceeded, and a difference in deflection, there was a vertical gap of 115 millimetres in mid-span between the two halves. There had been a similar problem also for the east span, however, due to a smaller distance the discrepancy was evened out by the help of jacks. A distance of 115 millimetres was too much for jacks to even out, so they had to come up with another idea this time. Unfortunately then somebody suggested that heavy concrete blocks were to be placed in mid-span on the girder half to be deflecting down (Fig. 15.14).

The extra strengthening of the free edge – that showed to be sufficient for loading coming from self-weight alone – showed to be *insufficient* for the additional loading coming from this temporary loading of the concrete blocks (the bending moment in

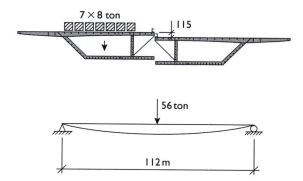

Fig. 15.14 An extra loading of 56 tons was put on one of the girder halves to even out the difference in deflection.

Fig. 15.15 The difference in deflection was evened out by the extra weight of the concrete blocks (to be seen on the left), but the free edge (of the extra loaded half) buckled, still being stiffened by the extra longitudinal stiffener at the free upper flange end. (Report of Royal Commission into the Failure of West Gate Bridge. Copyright State Government of Victoria, Australia)

mid-span increased some 15–20% relative to the bending moment of the self-weight alone). The difference in deflection evened out, however, the entire upper flange plate buckled (including the extra longitudinal stiffeners). In figure 15.15, the concrete blocks can be seen to the left, as well as the extra longitudinal stiffener at the free edge.

The continued assembly was taken to a halt completely, and for more than a month (!) they discussed about what to do about the situation. Finally it was decided to go ahead with the method used for the east span, i.e. to open up a transverse splice by removing the bolts. They started with the splice in mid-span and removed the bolts

Fig. 15.16 The "man-made" produced hinge made the bridge collapse to the ground.

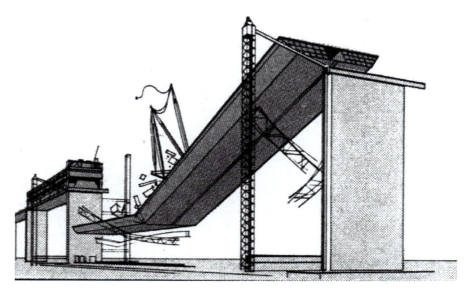

Fig. 15.17 A vivid drawing of the actual collapse. See also the back cover illustration. (Paul: När den stora bron rasade. Det Bästa 1972)

one by one – the difference though, in comparison to the east span, being that the buckle was much larger now and the loading heavier. As more and more bolts were removed, the stress increased on the remaining active part of the upper flange of the bridge – not only due to the loss of cross-sectional area, but also due to the gradual lowering of the neutral axis. After 16 bolts had been removed, there was a visible reduction of the buckles (through slipping of the plates), but there was also the effect that some of the remaining bolts were squeezed tight – because of the increased upper flange stress – which made the continued loosening more troublesome. When 37 bolts had been removed the inevitable collapse of the cross-section started – a buckle spread in the transverse direction (across the width of the bridge) due to the overloading of the net section, and the vertical web also buckled in its upper part (where it was subjected to compression). The remaining bolts of the upper flange sheared off due to the excessive shearing forces between the plates, and this was followed by a slow sinking of the left-hand bridge half (compare Fig. 15.15), which was now carried by the right-hand bridge half (the intact half) alone as there had been some connecting of the two halves earlier. However, as the loading of this right-hand bridge half also became too much for it to carry, it gave way for the excessive loading and the entire bridge fell some 50 metres to the ground – 36 people were killed on this tragic accident 15 October 1970. Just

Fig. 15.18 The collapsed girder span on the ground. (Report of Royal Commission into the Failure of West Gate Bridge. Copyright State Government of Victoria, Australia)

as the case was for the Cleddau Bridge in Milford Haven – where the buckling of the diaphragm produced a hinge that transformed the statically determinate system into a mechanism – this "man-made" hinge made the simply supported girder span of the West Gate Bridge collapse to the ground (the clearly visible hinge – as folding of the upper flange – can also be seen in the photo in Fig. 15.18) (Figs. 15.16, 15.17 and 15.18).

Every 16 metre one could see a local "ridge" in the upper flange, and this was the result of the weak detail described earlier of the splicing of the longitudinal stiffeners (can also be seen at the lower end in Fig. 15.18 above) (Fig. 15.19).

This local damage had nothing to do with the actual collapse of the entire bridge, instead it can be seen as a last and final "death-rattle".

Fig. 15.19 The local collapse of the deck plate due to the weak splicing of the longitudinal stiffeners (see also Fig. 15.11). (Report of Royal Commission into the Failure of West Gate Bridge. Copyright State Government of Victoria, Australia)

The collapse was caused by a series of faulty decision, which all started by the contractor choosing the rather unusual assembly method of having the girder in two separate halves. Stephenson did show already back in 1850, that it is not impossible to perform a lift of a 1600 ton heavy bridge section (which was the weight of the centre span tubes of the Britannia Bridge). If they had decided to carry out the lifting of the girder as a complete unit in the construction of the West Gate Bridge, the weight would "only"

Fig. 15.20 Some removed and saved parts from the collapsed West Gate Bridge at Monash University in Melbourne, Australia. (With kind permission of Bo Edlund)

have been 1200 tons, and then the problems concerning buckling and difference in vertical deflection would also have been gone.

The investigation stated – besides blaming the contractor – that Freeman, Fox & Partners had to take the major part of the responsibility:

- They had not checked the load-carrying capacity thoroughly enough for the assembly (and lifting) of the girder in two halves.
- During construction FF&P had been neglecting to answer questions regarding the structure as a whole.
- The inspection on site had been scarce, and carried out by a young and unexperienced civil engineer.
- Last, but not least, the main responsibility rested on FF&P for the fatal decision to remove the bolts.

If you ever visit The Monash University, Department of Civil Engineering, in Melbourne, then take the opportunity to study some of the remaining parts from The West Gate Bridge. Some "scrap pieces" have been saved there, outside the building, as a reminder of the importance of remembering bridge failures, and to learn from the mistakes made. On the left in the photo one can, for example, see the buckled deck plate at the longitudinal stiffener splice (Fig. 15.20).

Chapter 16

Rhine Bridge

On 10 November 1971, little more than a year after the collapse of the West Gate Bridge, a box-girder bridge over the River Rhine near Koblenz (Fig. 16.1), close to the inlet of the tributary River Mosel, fell down during construction, killing 13 workmen.

The Rhine Bridge (also called "The Südbrücke") was a continuous and haunched box-girder bridge in three spans, and would be one of West Germany's first ever all-welded bridges. The total length was 442 metres, having a centre span of 236 metres and two end spans of 103 metres each (Fig. 16.2).

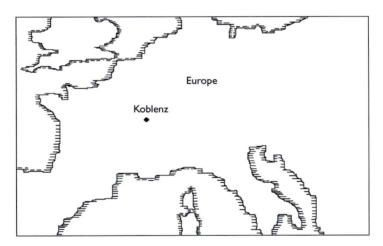

Fig. 16.1 The Rhine Bridge was located in Koblenz, in the southwestern part of former West Germany.

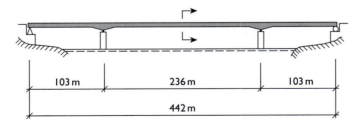

Fig. 16.2 Elevation of the Rhine Bridge.

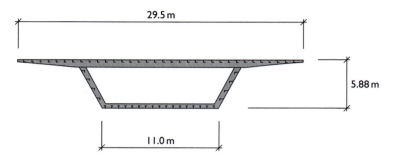

Fig. 16.3 Cross-section of the Rhine Bridge, near Koblenz.

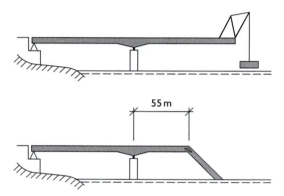

Fig. 16.4 During erection of the main span – using the free cantilevering technique – the cantilever arm buckled approximately half-way out and fell into the River Rhine.

The cross-section of the bridge in mid-span had an 11.0 metre wide lower flange, a 29.5 metre wide upper flange, and a depth of 5.88 metres. The upper flange and the inclined web plates were in the longitudinal direction strengthened by bulb-flat stiffeners (just as the case was for the Cleddau Bridge and the West Gate Bridge), while the lower flange had T-profiles (which, as a matter of fact, also was used for the Britannia Bridge) (Fig. 16.3).

In order to minimize the obstruction to boat traffic, it was chosen to cantilever out the centre span from both directions (just as for the Danube Bridge). The cantilever arm was about 100 metres long, and it was only the erection of the final section remaining (an 18 metre high lift of a 16 metre long and 85 ton bridge element from the water level), as the cantilever arm collapsed approximately halfway out (Figs. 16.4 and 16.5).

The negative bending moment in the cantilever arm, because of the loading coming from from the self-weight of the girder, the weight of the crane, and the new girder element (plus the additional inertia forces as the cantilever arm must be assumed to have been set in swaying), created too much compression for the lower flange to carry, which buckled and transformed the statically determinate system into a mechanism.

Fig. 16.5 Photo taken just after the collapse (see also Fig. 16.4).

The strange thing though, was that the buckle came halfway out, and not at the inner support, where the bending moment was the greatest. Several factors do, however, influence here:

- The bridge was haunched over the supports, i.e. stronger at these locations due to a larger construction depth.
- The position of the buckle – 55 metre out from the support – is approximately where one could expect the zero-moment points for the continuous system to be, and thus it can be assumed that the stiffeners in this position were "minimized" in size.
- Last, but not least, the buckle came exactly where the longitudinal stiffeners in the lower flange were spliced (i.e. in the joint between two girder elements) – a splice that quite clearly proved to be insufficient, as we will see in the following.

When two girder elements were to be joined together, there was a horizontal gap of 225 millimetres from each edge to allow for automatic welding in the transverse direction of the elements (without having to be interrupted by the many stiffeners) (Fig. 16.6).

As the longitudinal stiffeners were to be made intact again (i.e. continuous over the joint), the choice was to "hang" a T-profile on to the edges of two adjacent stiffeners, instead of filling the gap with an exact profile. The intention was (and quite rightly so, looking at the problem from one point of view) to avoid crossing welds which is a weakening factor concerning fatigue and brittle fracture strength (Fig. 16.7).

This choice of configuration created a free vertical distance of 25 millimetre between the inserted profile and the flange plate, which unfortunately did show to be the direct reason why buckling was initiated. Almost the same type of splice had been used at the West Gate Bridge (see Fig. 15.11), however, not being the "triggering factor" for the collapse (as it did show to be here).

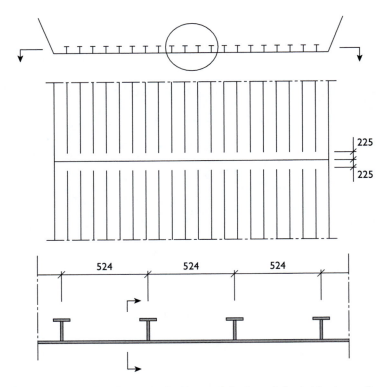

Fig. 16.6 The longitudinal stiffener configuration of the box-girder bridge lower flange.

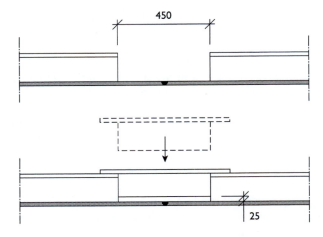

Fig. 16.7 The unfortunate choice of splicing longitudinal stiffeners.

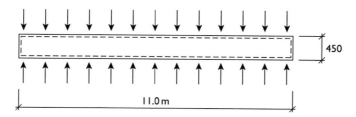

Fig. 16.8 The unstiffened part of the lower flange plate in between the longitudinal stiffener ends.

We will in the following study and analyze the load-carrying capacity of the lower flange plate in this more or less unstiffened zone. We lack information about the thickness of the plate and the yield strength of the steel material, however, as a qualified guess we assume $t = 10$ mm and $f_y = 360$ MPa. The flange plate is 11.0 metre wide and 450 millimetre long (in between the stiffener edges), and we assume simple supports along the four edges of the plate (Fig. 16.8).

We could perhaps assume the edges of this plate to a certain degree of being clamped to the stiffener ends, however, as these are widely separated (having a spacing of 524 mm), in relation to the distance in the longitudinal direction (i.e. 450 mm), we neglect this effect. The free buckling length of the flange plate is also longer (i.e. longer than 450 mm) in between the stiffeners.

We calculate the critical buckling stress for the flange plate (Eqs. 16.1 and 16.2):

$$k = \left(\frac{11.0}{0.450} + \frac{0.450}{11.0} \right)^2 = 599.5 \text{ (with } m = 1) \tag{16.1}$$

$$\sigma_{cr} = 599.5 \cdot \frac{\pi^2 \cdot 210000}{12 \cdot (1 - 0.3^2) \cdot \left(\frac{11.0}{0.010} \right)^2} = 94.0 \text{ MPa} \tag{16.2}$$

Just to assure ourselves of the reasonableness of this value, we calculate the critical *Euler* buckling stress (knowing that the difference in results, given the theory for a plate and the theory of a strut, should be small, as the loaded width is large) (Eq. 16.3):

$$\sigma_{Euler} = \frac{\pi^2 \cdot 210000}{12 \cdot (1 - 0.3^2) \cdot \left(\frac{450}{10} \right)^2} = 93.7 \text{ MPa} \tag{16.3}$$

The small difference in results – which confirms the validity – is also on the right side. The critical buckling stress, based on the plate theory, will always for very wide plates be slightly higher than the Euler buckling value, due to the double curvature at the edges.

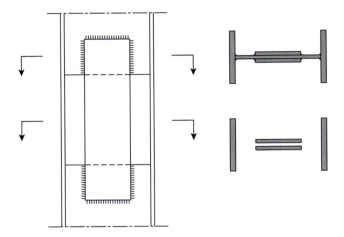

Fig. 16.9 A simplified model that could be used to explain and to understand the behaviour of the weak splicing of the longitudinal stiffeners in the box-girder bridge bottom flange – an axially loaded column having the web spliced with two cover plates over a free gap.

In order to receive the maximum load-carrying capacity in the ultimate loading state we calculate the effective width (according to the Eurocode) (Eq. 16.4):

$$\bar{\lambda}_p = \sqrt{\frac{360}{94.0}} = 1.957 \quad \rho = \frac{(1.957 - 0.22)}{1.957^2} = 0.454$$

$$\Rightarrow b_{eff} = 0.454 \cdot 11.0 = 4.99 \, \text{m} \tag{16.4}$$

We now compare this value with the effective width for the stiffened flange plate on either side of the spliced zone (i.e. the effective width for an 11.0 metre wide stiffened plate having 20 stiffeners, spaced 524 millimetres). As it is in this case a much smaller free distance in the transverse direction, the critical buckling stress will increase markedly (Eqs. 16.5 and 16.6):

$$\sigma_{cr} = 4.0 \cdot \frac{\pi^2 \cdot 210000}{12 \cdot (1 - 0.3^2) \cdot \left(\frac{524}{10}\right)^2} = 276.5 \, \text{MPa} \tag{16.5}$$

$$\bar{\lambda}_p = \sqrt{\frac{360}{276.5}} = 1.141 \quad \rho = \frac{(1.141 - 0.22)}{1.141^2} = 0.707$$

$$\Rightarrow b_{eff} = 0.707 \cdot 0.524 \cdot 21 = 7.78 \, \text{m} \tag{16.6}$$

In relation to the stiffened parts on either side, the capacity has been reduced to only 64% in the spliced zone (4.99/7.78) – a major decrease, that to an almost certainty was not taken into consideration.

As the flange plate was not in close (and composite) contact with the longitudinal stiffeners in the spliced zone, it buckled and thus triggered the collapse of the entire

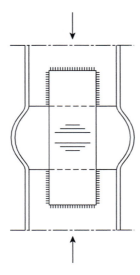

Fig. 16.10 Local buckling of the column having the web spliced.

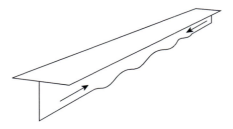

Fig. 16.11 Welding deformations in the transverse splice of the bottom flange plate tend to produce out-of-plane deformations.

Fig. 16.12 The free and unstiffened lower edge of the inserted T-profile (in the gap) was a weak part with respect to normal stress buckling.

cross-section. As a similitude one can consider a column, that has a very special splice where the web is missing, and that the web has been replaced with two separate cover plates (Fig. 16.9).

For such a weak splice it is not difficult to imagine the scenario at the first heavy loading – the flange plate of the column, and the cover plates, will for sure buckle out (as the splice more or less also can be regarded as an initial hinge) (Fig. 16.10).

Another factor which additionally weakened the lower flange plate splice was the inevitable welding deformation that forces the flange plate into a deflected shape in between the stiffener edges (Fig. 16.11, the deformation is heavily enlarged).

An additional effect, that also most certainly was not considered, was the weakness of the inserted T-profile, which had a an outstanding vertical part that was free and unsupported, and this made it prone for normal stress buckling along its lower edge (just as the case was for the free flange edge of the West Gate Bridge girder halves) (Fig. 16.12).

After the collapse, the Rhine Bridge was re-built, having the small gap between the splice profile and the flange plate closed, and with extra strengthening of the bridge in general. For the erection of the new bridge, free cantilevering was once again used (despite the consequence for the original bridge), but now with a reduced maximum cantilever length – a 60-metre long mid-section was this time lifted up from *two* shorter cantilever arms.

Chapter 17

Zeulenroda Bridge

Up until the end of the 1990's, the belief was that the four bridges in Vienna, Milford Haven, Melbourne, and Koblenz, were the only major examples of box-girder bridge collapses during erection. Imagine the surprise in 1998 as it was reported in the German scientific journal Stahlbau of another box-girder bridge collapse – a collapse that also had taken place in the early 1970's. The East German authorities had kept this incident secret to the Western world, and it was not until the archives were opened that the information became available to the general public. A probable reason why this was hidden away was the fact that the collapse happened, shamefully, at the same day as the anniversary of the Berlin Wall. Twelve year to the day after the building of the Berlin Wall, and less than four years after the buckling of the Danube Bridge, the box-girder bridge in Zeulenroda (approximately 100 kilometres south of Leipzig, close to the border of the Czech Republic, then Czechoslovakia – see Fig. 17.1) collapsed – 13 August 1973 – killing four people.

The bridge was to be a continuous box-girder bridge in six spans, having a total length of 362 metres. Just before the collapse, the second span was cantilevering half-way out, and the next element was to be assembled, the arm from then on to be carried by a temporary support in mid-span (Figs. 17.2 and 17.3).

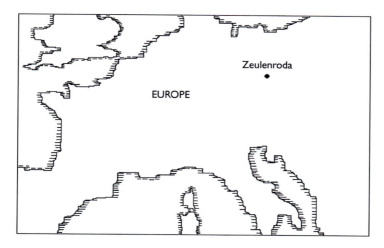

Fig. 17.1 The Zeulenroda Bridge was located in former East Germany.

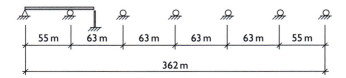

| 55 m | 63 m | 63 m | 63 m | 63 m | 55 m |

362 m

Fig. 17.2 Elevation of the Zeulenroda Bridge during erection.

Fig. 17.3 A longitudinal view of the Zeulenroda Bridge during erection. (Ekardt: Die Stauseebrücke Zeulenroda. Ein Schadensfall und seine Lehren für die Idee der Ingenieurverantwortung. Stahlbau. With kind permission of Karl-Eugen Kurrer)

However, the temporary support in mid-span never came to be in use, as the bridge suddenly gave way and fell to the ground (Figs. 17.4 and 17.5).

There was no exact and detailed explanation given in the Stahlbau article of the reason for the collapse, but still we will in the following give it a try to analyze the bridge cross-section to see if we can, somehow, prove that the bridge was inadequate in its design. The structural drawing, showing the cross-section of the box-girder, was the following (Fig. 17.6).

This drawing above gives us the information about the cross-section that we need for our analysis (Fig. 17.7).

The assembly of the bridge was, as has been mentioned earlier, at the stage where the cantilever was halfway out, and that the next element was to be connected to the cantilever end, from then on also supported by a temporary support (Fig. 17.8).

As the bridge gave way, the new element was not yet at the crane, instead it was positioned behind the same. The approximate positions of the crane and the element out on the cantilever were the following (Fig. 17.9).

In order to get an idea of the loading action on the cantilever arm at this particular moment during erection, we need not only the positions of the different loads, but also

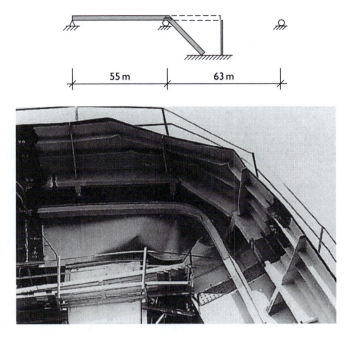

Fig. 17.4 Collapse mechanism and a close-up of the buckled corner (i.e. the hinge). (The photo: Ekardt: Die Stauseebrücke Zeulenroda. Ein Schadensfall und seine Lehren für die Idee der Ingenieurverantwortung. Stahlbau. With kind permission of Karl-Eugen Kurrer)

the size of the same (see Fig. 17.10). The weight Q of the crane, we set to an absolute minimum value:

$$Q \geq 100 \, \text{kN}$$

After analyzing the cross-section and finding the constants, we receive the remaining loads on the cantilever. The new element is 13.7 metres long, and is without transverse cantilevers:

$G = 305.8 \, \text{kN}$ (the weight of the new element)
$g = 25.8 \, \text{kN/m}$ (self-weight per metre bridge)
$D = 5.4 \, \text{kN}$ (the weight of the diaphragm)

The cross-sectional constants being:

$$A = 287976 \, \text{mm}^2$$
$$I_x = 19.362 \times 10^{10} \, \text{mm}^4$$
$$y_{\text{n.a.}} = 574 \, \text{mm}$$

Fig. 17.5 The unexpected and sudden collapse of the Zeulenroda Bridge. (Ekardt: Die Stauseebrücke Zeulenroda. Ein Schadensfall und seine Lehren für die Idee der Ingenieurverantwortung. Stahlbau. With kind permission of Karl-Eugen Kurrer)

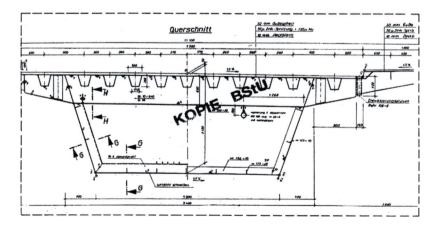

Fig. 17.6 Cross-section of the Zeulenroda Bridge, as given in the structural drawings. (Ekardt:
Die Stauseebrücke Zeulenroda. Ein Schadensfall und seine Lehren für die Idee der
Ingenieurverantwortung. Stahlbau. With kind permission of Karl-Eugen Kurrer)

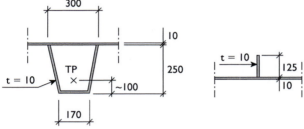

Fig. 17.7 The cross-section used for the analysis to come in this text.

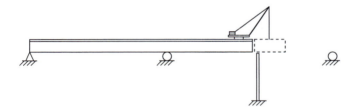

Fig. 17.8 The next sequence in the assembly of the bridge (if it not had been for the collapse that is).

The design bending moment for transient loading (i.e. temporary loading during construction) becomes (Eqs. 17.1 and 17.2):

$$M = 25.8 \cdot \frac{31.5^2}{2} + 5.4 \cdot (4.1 + 17.8 + 31.5)$$
$$+305.8 \cdot 12.1 + 100 \cdot (31.5 - 6) = 19339 \, \text{kNm} \qquad (17.1)$$

$$M_{sd} = \gamma_G \cdot M = 1.35 \cdot 19339 = 26107 \, \text{kNm} \qquad (17.2)$$

The maximum normal stresses in the upper and lower flanges become (Eqs. 17.3 and 17.4):

$$\sigma_{upper} = \frac{M_{sd}}{I_x} \cdot y_{n.a.} = \frac{26107 \cdot 10^{-3}}{19.362 \cdot 10^{-2}} \cdot 574 \cdot 10^{-3} = 77.4 \, \text{MPa} \qquad (17.3)$$

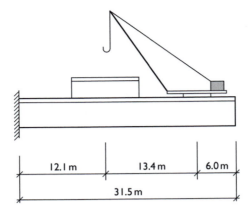

Fig. 17.9 The positions of the crane and the new element just prior to the collapse (see also Fig. 17.5).

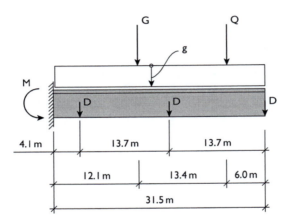

Fig. 17.10 The different loads acting on the cantilever arm.

$$\sigma_{lower} = \frac{M_{sd}}{I_x} \cdot (h - y_{n.a.}) = \frac{26107 \cdot 10^{-3}}{19.362 \cdot 10^{-2}} \cdot (2150 - 574) \cdot 10^{-3}$$
$$= 212.5\,\text{MPa} \tag{17.4}$$

The lower flange, that shall carry a compressive stress of 212.5 MPa, is stiffened in the longitudinal direction by five flat steel bars and in the transverse direction by a stiffener each 2.74 metres (Fig. 17.11).

The first step will be to check the maximum capacity of the flange plate *in between* the longitudinal stiffeners. We start by calculating the critical buckling stress of the same. We assume the free width of being equal to the centre distance between the stiffeners (Eqs. 17.5 and 17.6):

$$a = 2.74\,\text{m} \quad b = \frac{4.0}{6} = 0.667\,\text{m}$$

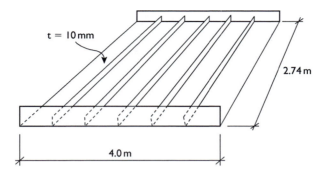

Fig. 17.11 Transverse and longitudinal stiffeners in the lower flange.

$$\frac{a}{b} = \frac{2.74}{0.667} = 4.1 \Rightarrow k = 4 \tag{17.5}$$

$$\sigma_{cr} = 4 \cdot \frac{\pi^2 \cdot 210000}{12 \cdot (1 - 0.3^2) \cdot \left(\frac{667}{10}\right)^2} = 170.6 \, \text{MPa} \tag{17.6}$$

Already here we see that something is not quite right – the actual compressive stress of 212.5 MPa is markedly exceeding the ideal critical buckling stress – and we can suspect that the post-critical reserve strength is not going to be enough. However, we will still carry on with the analysis. First we check the cross-section class of the flange plate. The quality of the steel is St 38, which has a yield strength of 235 MPa (Eqs. 17.7 and 17.8):

$$\text{Class 3:} \quad \frac{b}{t} \le 42 \cdot \sqrt{\frac{235}{f_y}} = 42 \tag{17.7}$$

$$\text{Actual slenderness:} \quad \frac{b}{t_f} = \frac{667}{10} = 66.7 \tag{17.8}$$

$$\Rightarrow \text{Class 4}$$

We continue by calculating the effective width (for a cross-section in Class 4) (Eqs. 17.9–17.11):

$$\overline{\lambda}_p = \sqrt{\frac{235}{170.6}} = 1.173 \tag{17.9}$$

$$\rho = \frac{(1.173 - 0.22)}{1.173^2} = 0.693 \tag{17.10}$$

$$\Rightarrow b_{eff} = 0.693 \cdot 667 = 462 \, \text{mm} \tag{17.11}$$

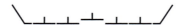

Fig. 17.12 As a simplified approach the longitudinal stiffeners are assumed to buckle independently of each other.

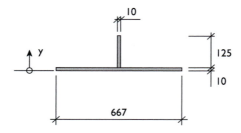

Fig. 17.13 Cross-section of the longitudinal stiffener and the interacting part of the lower flange plate.

The load-carrying capacity – expressed as the axial normal force capacity – then becomes (Eq. 17.12):

$$N_{c.Rd} = \frac{A_{eff} \cdot f_y}{\gamma_{M1}}$$
$$= \frac{462 \cdot 10^{-3} \cdot 10 \cdot 10^{-3} \cdot 235 \cdot 10^3}{1.0} = 1086 \, \text{kN} \tag{17.12}$$

which we compare to the design axial normal force (Eq. 17.13):

$$N_{c.Sd} = A_{tot} \cdot \sigma_{lower}$$
$$= 667 \cdot 10^{-3} \cdot 10 \cdot 10^{-3} \cdot 212.5 \cdot 10^3 = 1417 \, \text{kN} \tag{17.13}$$
$$\Rightarrow \text{Insufficient load-carrying capacity!}$$

If the flange plate (in between the stiffeners) would have had a *sufficient* load-carrying capacity, the next step in our analysis would have been focusing on the longitudinal stiffeners alone. For the completeness of our analysis we continue to do this check, and we assume the longitudinal stiffeners are having the dimension that was originally given in the cross-section, i.e. $125 \times 10 \, \text{mm}^2$ – the thickness is, however, crossed out and adjusted to 20 millimetres (an adjustment made afterwards, as a correction of what it should have been?). The simplest model, but also the most practical one, is to regard the longitudinal stiffeners as separate units, without any support in the transverse direction from the flange plate. The stiffeners are then assumed to buckle independently of each other, in the out-of-plane direction relative the flange plate plane (in between the transverse cross-beams) as *isolated* elements (Fig. 17.12).

We then have the following cross-section to analyze (Fig. 17.13).

A check of the cross-section class of the longitudinal stiffener (Eqs. 17.14 and 17.15):

$$\text{Class 3:} \quad \frac{c}{t} \le 14 \cdot \sqrt{\frac{235}{f_y}} = 14 \tag{17.14}$$

$$\text{Actual slenderness:} \quad \frac{c}{t} = \frac{125}{10} = 12.5 \tag{17.15}$$

As the actual slenderness lies below the slenderness limit of Class 3, the longitudinal stiffener (being subjected to axial compression) has a full capacity to fully plastify. Note that this is only with respect to the *local* buckling risk.

Cross-sectional constants:

$$A_{gross} = 7920 \, \text{mm}^2$$

$$A_{eff} = 5870 \, \text{mm}^2 \qquad \text{(with the effective width as before)}$$

$$y_{n.a.} = 15.7 \, \text{mm}$$

$$I_x = 6.48 \times 10^6 \, \text{mm}^4$$

The design buckling resistance (Eqs. 17.16–17.23):

$$N_{b.Rd} = \frac{\chi \cdot \beta_A \cdot A \cdot f_y}{\gamma_{M1}} \tag{17.16}$$

$$i = \sqrt{\frac{I}{A}} = \sqrt{\frac{6.48 \cdot 10^6}{7920}} = 28.6 \, \text{mm} \tag{17.17}$$

$$\lambda = \frac{l_c}{i} = \frac{2.74}{28.6 \cdot 10^{-3}} = 95.8 \tag{17.18}$$

$$\lambda_1 = 93.9 \cdot \varepsilon = 93.9 \tag{17.19}$$

$$\beta_A = \frac{A_{eff}}{A} = \frac{5870}{7920} = 0.741 \tag{17.20}$$

$$\bar{\lambda} = \frac{\lambda}{\lambda_1} \cdot \sqrt{\beta_A} = \frac{95.8}{93.9} \cdot \sqrt{0.741} = 0.878 \tag{17.21}$$

$$\text{buckling curve c (welded)} \quad \Rightarrow \quad \chi = 0.614 \tag{17.22}$$

$$N_{b.Rd} = \frac{0.614 \cdot 0.741 \cdot 7920 \cdot 10^{-6} \cdot 235 \cdot 10^3}{1.0} = 847 \, \text{kN} \tag{17.23}$$

To be compared to the actual normal force that a longitudinal stiffener shall carry (we assume the normal stress to be evenly distributed over the cross-section) (Eq. 17.24):

$$N_{b.Sd} \approx 212.5 \cdot 10^3 \cdot 7920 \cdot 10^{-6} = 1683 \, \text{kN} \tag{17.24}$$

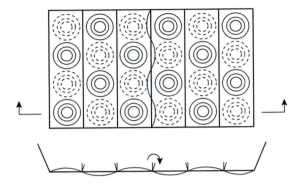

Fig. 17.14 The sine-wave buckling pattern in the flange plate has the effect on the longitudinal stiffeners that they tend to rotate in a sine-wave pattern as well.

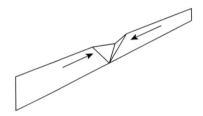

Fig. 17.15 Possible "folding" mechanism of a longitudinal stiffener due to a local damage.

From the results we see that the longitudinal stiffeners also were given too small dimensions, so the bridge was quite clearly doomed to fail during erection. And even if consideration is made regarding a more refined global buckling model – where for example the influence of the stiffening effect from the flange plate in the transverse direction is taken into consideration – there are more reducing effects that have to be considered which *lower* the load-carrying capacity. In order for a longitudinal stiffener to function as a rigid nodal point at buckling of the flange plate, it has not only to be sufficiently stiff in the out-of-plane direction (i.e. in the vertical direction relative the plane flange plate), but also be torsional stiff in order not to deform in the transverse direction due to the plate buckling (Fig. 17.14).

Flat steel bars as stiffeners are "torsional soft", which make them susceptible to welding deformations, and this increases the tendency even more for lateral deformations. In addition, there is also the probability of hits and damages during transport, handling, and assembly. A local damage can make the stiffener lose most of its load-carrying capacity with respect to the global buckling resistance (Fig. 17.15).

And if one longitudinal stiffener is eliminated, it also means that the total load-carrying capacity of the flange plate is reduced (as the effective width of the plate decreases) (Fig. 17.16).

Thus, the advice is never to choose flat bars as longitudinal stiffeners (as was done not only in the Zeulenroda Bridge, but also in the Fourth Danube Bridge, see Chapter 12) because of the above-mentioned reasons. Besides a negligible rotational (torsional)

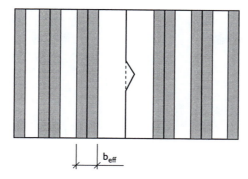

b_{eff}

Fig. 17.16 The negative effect on the overall ultimate load-carrying capacity of a stiffened plate because of a locally damaged stiffener.

Fig. 17.17 A more desirable choice of stiffener types.

stiffness, these profiles have not enough bending stiffness in the transverse (horizontal) direction. Profiles having a large bending stiffness in the transverse direction (besides the vertical direction) are for example L-profiles or T-bars (Fig. 17.17).

There are also more advantages than have been discussed above using these kinds of profiles in comparison to flat steel bars. A flat steel plate has an outstand part, that is, even if it is straight and without imperfections, very sensitive for *normal stress buckling* (see also the discussions regarding the West Gate Bridge and the Rhine Bridge). The relatively higher torsional stiffness of L- or T-profiles gives also a *higher clamped condition* for the flange plate in these positions (the transverse rotation is prevented to a higher degree – see Figs. 17.14 and 17.17), which increases the critical buckling stress of the flange plate.

As a final comment regarding the choice of longitudinal stiffener cross-section for the Zeulenroda Bridge one can say that the wrong profile was chosen, in combination with an insufficient load-carrying capacity of the bridge as a whole. Besides a thicker flange plate, and more rigid stiffeners, it would also have been motivated with a greater depth of the bridge cross-section that would have reduced the bending stresses. The girder depth of the Zeulenroda Bridge was very small in relation to the span length (compare the cross-section for example to that of the Cleddau Bridge, see Figs. 17.7 and 14.3).

Chapter 18

Reichsbrücke

When it was decided to replace the old truss bridge (Fig. 18.1) over the River Danube in Vienna, Austria (for map see Fig. 12.1), back in the 1930's, the choice fell upon a suspension bridge (Fig. 18.2).

The name of the old bridge had in 1919 been changed to the Reichsbrücke (the "Empire Bridge") in the aftermath of the First World War, when Austria became a republic (it was the presumptive heir to the Austrian/Hungarian throne – Archduke Franz Ferdinand – that was assassinated in Sarajevo in 1914 which had triggered the First World War, as a matter of fact). During the construction of the new suspension bridge the old bridge was continuously in use, resting on temporary supports downstream of the old location (see Fig. 18.1 which most probably shows the extended

Fig. 18.1 The old truss bridge over the River Danube in Vienna – Kronprinz-Rudolph-Brücke – which was constructed in 1876. (www.reichsbruecke.net/geschichte_e.php)

Fig. 18.2 The new bridge was opened in 1937, and replaced the old truss bridge. (www. reichsbruecke.net/geschichte_e.php)

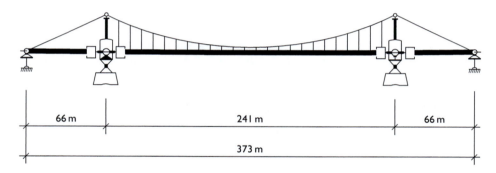

Fig. 18.3 Elevation of the Reichsbrücke. (first published by IABSE in Reiffenstuhl, H.: Collapse of the Viennese Reichsbrücke: Causes and Lessons. IABSE Symposium, Washington 1982)

bridge piers just before the sliding operation). The original piers were used for the new suspension bridge.

The new bridge carried four lanes, two tram lines and two footpaths, and the structure itself was an eye-bar chain suspension bridge, similar in type to the Point Pleasant Bridge (see Chapter 11), however, not having the main cable integrated with a stiffening truss (i.e. not being of the Steinman type). Otherwise it was identical with respect to the rocker towers, which are needed when the main cable is not continuous over the tower saddles, as is the case when using eye-bar chains. With its main span of 241 metres it was the third longest eye-bar chain suspension bridge in Europe at that time (Fig. 18.3).

The main cable was not anchored in any of the ordinary ways that is the case for suspension bridges (to solid rock or by a gravity anchor), instead a self-balancing

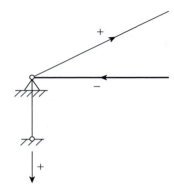

Fig. 18.4 A self-balancing system was chosen for the anchorage of the cable, most probably because of poor soil conditions.

system was used where the main cable was anchored to the end spans (where the cable bent normally is located) having the horizontal pull of the cable force balanced by compression in the stiffening girder (Fig. 18.4).

During the Second World War the Reichsbrücke was the only bridge over the Danube River in Vienna that survived from being destroyed. In the period between 1946–1953 after the war it was named Brücke der Roten Armee (the Red Army Bridge) being located in the Soviet Union zone as it was (the City of Vienna was divided between and controlled by the Allied forces – the Soviet Union, Britain, France and USA).

For almost 40 years the Reichsbrücke served its purpose well, but early one Sunday morning, at 04:45 on 1 August 1976 – little less than seven years after the collapse of the Fourth Danube Bridge in Vienna (see Chapter 12) and almost exactly 100 years after the opening of the old Kronprinz-Rudolph-Brücke (taken into service 21 August 1876) – the bridge collapsed without warning and fell into the Danube River. Due to the early hours only four vehicles (including a bus) were on the bridge, and only one driver of a passenger car lost his life. Five pedestrians ran for their lives and successfully reached safe ground before the bridge fell down (Fig. 18.5).

At first the authorities suspected a terrorist sabotage and had the other Danube bridges in Vienna guarded by police – a suspicion which is quite understandable as the collapse had occurred when there was more or less no traffic load on the bridge – but soon it became evident that the collapse had been initiated by the failure of the river pier (Fig. 18.6).

An expert commission was summoned to investigate the reason behind the collapse, and after a few months they presented their results. It was confirmed that the failure was due to the fracturing of the river pier, consisting of unreinforced concrete as it was. The effects of creep and shrinkage were also said to be contributing factors. The failure could not have been predicted the commission said. Knowing the fact though that the bridge piers were *unreinforced* indicates a long-term deterioration because of a gradual change of the inner stress state of the same. We will in the following consider a probable failure sequence.

Fig. 18.5 The collapsed Reichsbrücke. (www.reichsbruecke.net/geschichte_e.php)

Fig. 18.6 The end span supported by the collapsed river pier (to the bottom right in Fig. 18.5 above). (www.reichsbruecke.net/geschichte_e.php)

The vertical load from the bridge tower is distributed and dispersed into the pier (Fig. 18.7).

The inclined struts (especially the outer) have to be balanced in the lower part by a horizontal force (tie) which is introducing tension to concrete core. However, as the balancing force is small (at least at an initial stage) – due to the small inclination

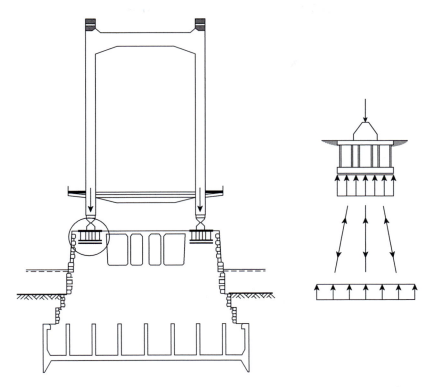

Fig. 18.7 Cross-section and load distribution and dispersion under the tower foot. (Left-hand figure: first published by IABSE in Reiffenstuhl, H.: Collapse of the Viennese Reichsbrücke: Causes and Lessons. IABSE Symposium, Washington 1982)

of the inclined struts – the unreinforced concrete is able to carry this tensile loading (Fig. 18.8).

As the time-dependant phenomenon of creep makes the stiffness of the concrete to decrease over time, more and more load is gradually transferred to the granite stone wall cladding (Fig. 18.7) – which is, in contrast to concrete, not susceptible to creep – thus increasing the inclination of the outer strut (in our strut-and-tie model), and thereby also increasing the need for a balancing horizontal tensile force (Fig. 18.9).

When the tensile strength of the concrete is exceeded a vertical crack, perpendicular to the applied tension, is formed in the interior of the bridge pier – a failure mode which goes by the name of splitting. The inclined strut has now to be balanced by horizontal *compression* exerted by the granite wall (Fig. 18.10).

However, as the strength of the granite wall in the horizontal direction is more or less negligible, the load is instead transferred to the remaining two intact struts – increasing the vertical pressure, but also shifting the need for balancing forces. The inclined inner strut now needs to be balanced by horizontal *tension* in the upper part, instead of

Fig. 18.8 Balancing tension (as well as balancing compression) in a strut-and-tie model.

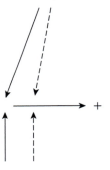

Fig. 18.9 The gradual loss of stiffness of the concrete (under constant loading) transfers more and more load towards the stiffer exterior.

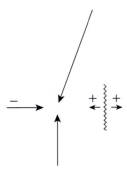

Fig. 18.10 Vertical cracking (splitting) increases the horizontal pressure on the granite wall.

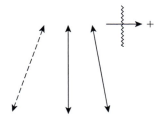

Fig. 18.11 Splitting failure in the upper part following cracking in the lower.

compression as it was before, and this shift (precipitated by the cracking in the lower part for the outer strut) leads to splitting failure in the upper part as well (Fig. 18.11).

The only load-carrying portion that now remains is a very unstable central vertical core which is being crushed by the pressure from the tower foot, and the final failure is where large chunks of concrete are being spalled away, leaving the tower foot unsupported (being equal to the final collapse of the bridge) (Fig. 18.12).

A heat wave followed by a sudden change to very cold weather at the time just before the collapse could also have been a contributing factor, as the contraction of the exterior parts of the pier was restricted by the still warm interior, leading to additional tensile stresses. However, the influence of these temperature stresses is not significant – this additional and possibly contributing effect can be seen as a triggering factor only. Back in 1936 though, when the bridge was under construction, it has been reported that one of the old bridge piers was hit by a ship, possibly the same one that was used for the suspension bridge – see Figures 18.1 and 18.2 – and if so, this would most certainly have created some damage to the pier. Whether this contributed to a loss of strength of the pier can only be guessed.

Back in the 1930's creep and the possibility of splitting failure was not known phenomena, but the designers should anyway have been well aware of the low tensile strength of unreinforced concrete and added some minimum reinforcing bars locally under the tower feet. Some horizontal stirrups would readily have carried the stabilizing force (i.e. acting as a strong tie) (Fig. 18.13).

Another measure, that perhaps would have saved the bridge, if taken, would have been a closer and more detailed inspection of the substructure. Less thorough inspections tend to concentrate on the superstructure, and on details and structural members that are easy accessible. The Point Pleasant Bridge failed because of lack of maintenance and inspections "up there" (see Chapter 11), and the Reichsbrücke because of lack of inspections "down there" – locations that are equally inaccessible. And speaking about the Point Pleasant Bridge – which was an eye-bar chain suspension bridge just like the Reichsbrücke – it also failed just before reaching 40 years in service; perhaps that is the life-time to be expected of these kinds of bridges?

And yet another similarity to other bridges that have failed; just like the Tay Bridge, where the engine was saved and later put in use (see Chapter 3, Fig. 3.22), the bus on the Reichsbrücke the morning the bridge collapsed was salvaged, repaired and again taken into service (whether or not the bus driver continued to drive the same bus the story does not tell) (Fig. 18.14).

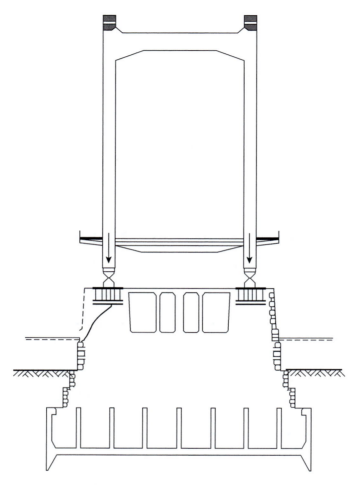

Fig. 18.12 When the tower leg lost its footing the bridge collapsed, tilting to the left. (first published by IABSE in Reiffenstuhl, H.: Collapse of the Viennese Reichsbrücke: Causes and Lessons. IABSE Symposium, Washington 1982)

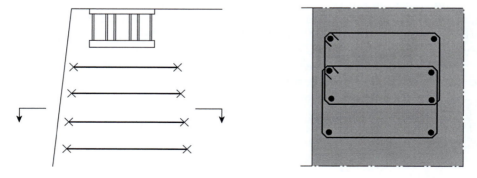

Fig. 18.13 Configuration and position of horizontal stirrups that would have prevented the collapse. (The vertical bars are not shown in the figure to the left)

Fig. 18.14 The bus that was salvaged from the Danube River. (www.wien-vienna.at/geschichte. php?ID=737)

Fig. 18.15 A concrete twin-box-girder bridge was opened in November 1980. This bridge is also called the Reichsbrücke. (www.reichsbruecke.net/geschichte_e.php)

During planning and construction of a new bridge, two military truss bridges were in the mean time rapidly assembled to span the Danube River in order to accommodate for the traffic. In November 1980, four years after the collapse, a continuous prestressed concrete twin-box-girder bridge was completed and taken into service (Fig. 18.15).

Chapter 19

Almö Bridge

When it was decided, back in the 1950's, to connect the mainland to the island of Tjörn at Stenungsund, some 45 kilometres north of Göteborg (Gothenburg) in Sweden, the choice fell upon a tubular steel arch bridge to span the Askerö Sound between the small islands of Källön and Almön (two other bridges were also part of this Tjörn Bridge project) (Fig. 19.1, see also Fig. 19.5).

With its main span of 278 metres the Almö Bridge would not only become the only bridge in Sweden of this type, but also the longest one – breaking the record of the Sandö Bridge (see Chapter 6) – and would also become the longest tubular steel arch bridge in the world. This particular bridge type was chosen because of aesthetical reasons, as it would harmonize with the archipelago landscape of the province Bohuslän – a suspension bridge was said not to be an alternative as it would stand out with its high pylons, and additionally being a risk of getting hit by aircrafts. However, it would show that the bridge was eventually brought down all the same, but being hit by quite a different type of traffic.

The construction of the bridge began in 1956, and the erection method chosen was to use temporary towers which carried a cable to which the tubular arch was suspended (Fig. 19.2).

The bridge was opened in 1960 and consisted of two parallel tubes of 3.8 metres in diameter and with a thickness ranging from 14 mm (at the crown) to 22 mm (at the base) (Figs. 19.3 and 19.4). In the load application points – i.e. where the vertical columns were applied – the tubes were ring-stiffened. No longitudinal stiffeners were used due to the inherent stability of the optimal circular cross-section (we will come to that later on). The tubes were braced (cross-framed) in the lateral direction by a lattice/girder system.

The Askerö Sound was the only available route for large ships going to the seaport of Uddevalla (some 35 kilometres to the north of Stenungsund), and the passage was through the narrow gap at the Almö Bridge (Fig. 19.5).

Through the bridge the free sailing width – with a maximum sailing height of 41 metres – was only 50 metres (marked by special signalling lamps attached to the arches) (Fig. 19.6).

And every once in a while very large tankers, like the "Thorshammer" of 228,000 tons, were allowed to pass the bridge (guided by a pilot) (Fig. 19.7).

Images like the "Thorshammer" makes one wonder if the designers had suppressed the fact that sooner or later a ship has to come off course and hit the bridge, especially

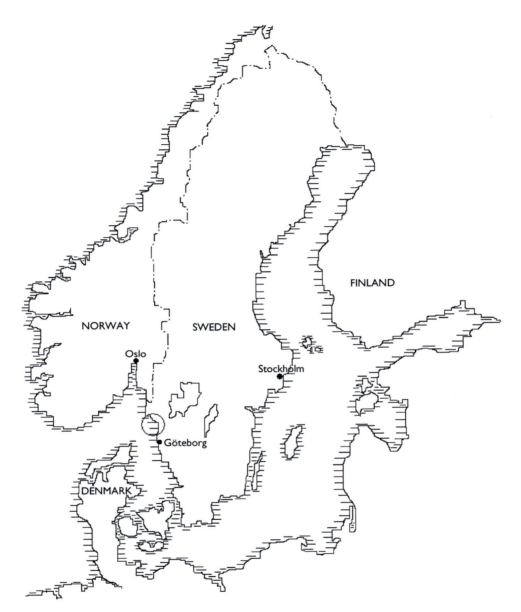

Fig. 19.1 The Almö Bridge was located at the West Coast of Sweden, just north of Göteborg.

as no barricades and warning devices (besides the signalling lights) had been provided. It was just a matter of time – little short of 20 years to be exact – before the inevitable thing happened. On the dark and misty night of 18 January 1980, at about 01.30, the cargo ship "Star Clipper" (27,000 tons, but not loaded to the full), heading to

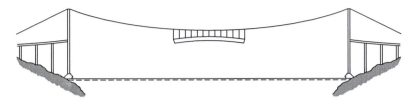

Fig. 19.2 Temporary erection towers (pylons) during the construction of the bridge. This particular erection method was chosen not only because of the great depth of the Askerö
Sound (some 40 metres), but it also gave free way for the ships (see Fig. 6.10 for the
contrast).

Fig. 19.3 The Almö Bridge as completed.

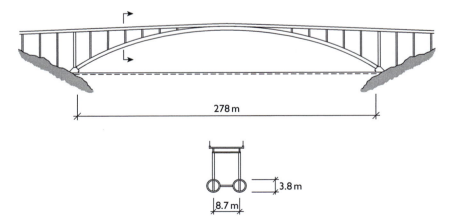

278 m

3.8 m

8.7 m

Fig. 19.4 Elevation and cross-section.

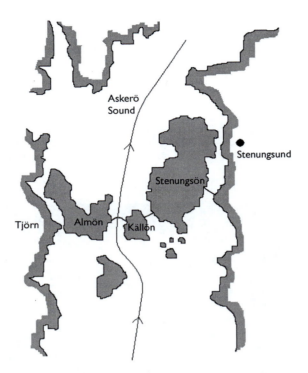

Fig. 19.5 The navigation route through the Askerö Sound to the seaport of Uddevalla.

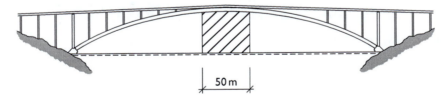

Fig. 19.6 With a sailing height of maximum allowed 41 metres the free sailing width was limited to 50 metres only.

Uddevalla, hit the bridge only some 35 metres from the west abutment (i.e. on the Almö side) (Fig. 19.8).

To the despair of the ship's crew seven cars went over the edge before the traffic was stopped – the personnel tried in vain to warn the motorists by shooting signal rockets. Eight people died, and no one survived the fall into the water.

In the following investigation by the governmental commission much criticism was directed to the pilot – for not having awaited daylight and using a tug boat because of the mist – to the Swedish Maritime Administration – for not having provided for better rules regarding large ships in narrow channels – and to the rescue work in general – for taking so long before the bridge was closed to traffic. But what was not discussed

Fig. 19.7 The passage of the giant tanker "Thorshammer" on 6 December 1969 – length 325 metres and width 48 metres. The clearance at the top was only 1.5 metres. Something like doing the Limbo Dance!

Fig. 19.8 The collapse of the Almö Bridge. (Tjörnbron. The Swedish National Road Administration. With kind permission of Lennart Lindblad)

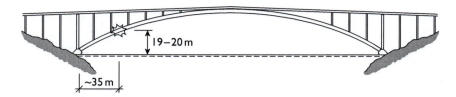

Fig. 19.9 The location where the bridge was hit by the Star Clipper gantry crane.

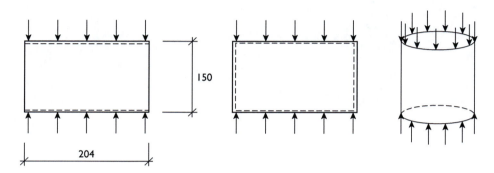

Fig. 19.10 A flat aluminium foil (204 × 150 × 0.18 mm³) as an axially loaded *strut* (supported at the loaded edges only), a *plate* (supported at all four edges), and a *cylinder* (where the foil has been folded together) having a diameter of 65 millimetres (not in scale).

was whether this was an acceptable construction with respect to the probability of being hit by ships coming off course. You should, as a designer, *always* consider the possibility of hit damages, and seen from this perspective this particular bridge should *never* have been built!

Despite being susceptible to ship collisions this particular bridge type was also very vulnerable to hit damages, even very small ones – something that most probably was not taken into consideration in the design process as well.

The ship hit the bridge tubes with its overhead gantry crane at a height of approximately 19–20 metres above the water level (some 35 metres from the shore) (Fig. 19.9).

Whether or not the blow from the crane can be regarded as "small" cannot be ascertained for sure, but in the following discussion we assume that that was the case (but with respect to the stability of thin-walled tubes – carrying large normal forces – it does not matter if the horizontal blow is large or small, as it will collapse anyhow).

Circular tubes (cylinders) are the most optimal shape to carry axial loading with respect to the amount of material used (being equally as strong with respect to global buckling in any direction), but are at the same time very sensitive to local disturbances (even small ones) that affect the equilibrium of the stabilizing membrane forces. Consider the following comparison between a strut, a plate, and a tube, all made up of one and the same aluminium foil (thickness $t = 0.18$ mm). We calculate the critical buckling stress for these three different cases (Fig. 19.10).

The critical buckling stress for the first alternative (the foil acting as a *strut*) (Eq. 19.1):

$$\sigma_{cr} = \frac{\pi^2 \cdot E}{12 \cdot (1 - \upsilon^2) \cdot \left(\dfrac{150}{0.18}\right)^2} = 0.09 \text{ MPa} \tag{19.1}$$

(with $E = 70{,}000$ MPa and $\upsilon = 0.33$ for aluminium)

The critical plate buckling stress for the second alternative (Eq. 19.2):

$$\sigma_{cr} = k \cdot \frac{\pi^2 \cdot E}{12 \cdot (1 - \upsilon^2) \cdot \left(\dfrac{204}{0.18}\right)^2} = 0.22 \text{ MPa} \tag{19.2}$$

(with $k = \left(\frac{204}{150} + \frac{150}{204}\right)^2 = 4.39$, and the plate *width* instead of the depth in the denominator)

The critical buckling stress for the third alternative (the cylinder) (Eq. 19.3):

$$\sigma_{cr} = \frac{1}{\sqrt{3 \cdot (1 - \upsilon^2)}} \cdot E \cdot \frac{0.18}{\left(\dfrac{65}{2}\right)} = 237 \text{ MPa} \tag{19.3}$$

If we compare the results between the cylinder and the plate – as these two are examples of *local* buckling – we find that the critical buckling stress has increased more than thousand times (237/0.22 = 1077)! This comparison is, however, somewhat false, as there is a post-critical reserve strength to consider for the plate in the ultimate limit state design, and that the theory for shell buckling is heavily overestimating the true load-carrying capacity (based on pure ideal conditions and not taking the influence of imperfections into account). In our case, where we are studying an aluminium plate, we have to consider the low yield stress level of the same as well (which is about 30 MPa or so, i.e. much lower than the critical shell buckling stress). But still, the comparison shows the huge advantage of using the material in the most optimum shape possible.

The secret behind the optimal behaviour of an axially loaded shell is that it develops membrane forces that keep the wall in balance – any tendency for outward or inward buckling is counterbalanced by the initiation of stabilizing forces in the circumference direction (Fig. 19.11). One could say that the shape is stabilizing itself, whereas plated bridge members would need to be strengthened by the use of longitudinal stiffeners (which, in the use of cylinders, can be omitted).

The *outward* going buckling tendency is balanced (held back) by tension in the membrane, just like the action of a suspension bridge (to make a similitude) (Fig. 19.12).

And, similarly, the *inward* going buckling tendency is balanced (held back) by compression in the membrane, just like the action of an arch bridge (Fig. 19.13).

However, this equilibrium can easily be disturbed by the presence of initial imperfections or by the initiation of local damages. Especially inwards going deformations have a tendency to increase in size due to the membrane (compression) forces, which could initiate a local failure, which is then equal to the entire collapse of the cylinder as the equilibrium is lost (Fig. 19.14).

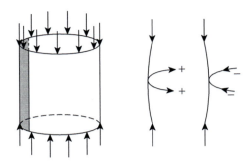

Fig. 19.11 Stabilizing membrane forces keep the wall in balance.

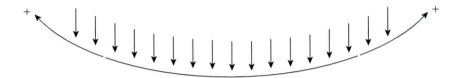

Fig. 19.12 Membrane forces in tension, similar in action to that of a suspension bridge. Also applicable for the case if the cylinder would have been subjected to inside pressure.

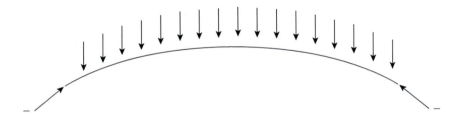

Fig. 19.13 Membrane forces in compression, similar in action to that of an arch bridge. Also applicable for the case if the cylinder would have been subjected to outside pressure (or inside suction).

If we now go back to the aluminium cylinder that we studied before, it so happens that it has exactly the same dimensions as one particular type of beer can (see once again Fig. 19.10), and we now take the opportunity to find its design load-carrying capacity for axial loading. We must consider the fact that the yield stress was much lower than the critical buckling stress, as well as factors such as production method, tolerance class (taking initial imperfections into account) and, above all, the so-called knock-down factor. The knock-down factor is the influence from initial imperfections on the ability of the membrane forces to stabilize the cylinder wall – as the slenderness ratio, r/t, increases the knock-down factor decreases (read: the knock-down effect on the stability increases). It could also be expressed like this; as the curvature decreases the membrane forces become more and more susceptible to local deformations, thus more easily disturbed (the extreme being when the curvature is zero – i.e. when we

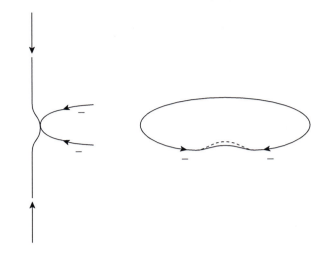

Fig. 19.14 Inwards going deformations tend to increase in size as they are "pushed on" by the membrane forces (in compression). The opposite applies for outwards going deformations as they instead tend to be "held back" by the membrane forces (in tension).

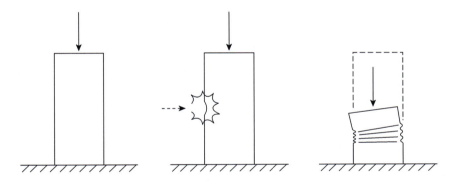

Fig. 19.15 If a small horizontal blow is applied at the side of the aluminium can – at the same time it is axially loaded with a large normal force – there will be an immediate collapse.

have a flat plate – then, as a consequence, there cannot be any membrane forces to balance a local damage prior to buckling).

The design strength resistance ends up being just below 20 MPa, giving a load-carrying capacity of about 70 kg. This result could be confirmed by asking a person of that particular weight to stand on an identical can, and we could with our own eyes see that the load-carrying capacity is sufficient. Knowing that this particular loading is just about what the aluminium can is able to carry we also know that the equilibrium is fragile (read: the membrane forces can easily disturbed as the axial loading is high). If we would ask the person to remain standing on the can we could take a ruler (or something similar) and just tap very lightly at the can wall – the can will then immediately collapse, as we have destroyed the stabilizing function of the membrane (Fig. 19.15).

But then the Almö Bridge tubular arch is not a "beer can" – with respect to material (steel instead of aluminium) and slenderness ratio – or is it? It is true that steel material have a much larger modulus of elasticity, E, being three times as high (210,000/70,000), which will increase the critical buckling stress (see Eq. 19.3), and a very much higher yield strength that gives a higher design strength, but with respect the slenderness of the tube wall the difference is not that dramatic. We compare the slenderness ratio of the maximum value for the Almö Bridge (which is at the crown) to that of the aluminium can (Eqs. 19.4 and 19.5).

$$\left(\frac{r}{t}\right)_{Alm\ddot{o}} = \frac{1.9}{0.014} = 136 \tag{19.4}$$

$$\left(\frac{r}{t}\right)_{can} = \frac{32.5}{0.18} = 181 \tag{19.5}$$

We see that the difference is not that big, and as a matter of fact, if the minimum thickness of the tube wall at the Almö Bridge would have been 10.5 millimetres (instead of 14) then the slenderness ratios would have been exactly the same ($1.9/0.0105 = 181$)! It is true, however, that the utilization with respect to the maximum load-carrying capacity was far from similar – the aluminium can being loaded to its maximum (70 kg), while the arches were only carrying the self-weight of the bridge at the time of the collapse (some 65 MPa at the crown, and approximately 50 MPa at the base – however, equivalent to a normal force of more than 1000 tons!). So it would require much more than a light tap by a "scaled-up ruler" to make the Almö Bridge collapse, but the discussion above anyway shows that the failure of the bridge very well could have been by local shell buckling initiated by a relatively small horizontal blow from the gantry crane.

Perhaps it is somewhat rude to compare the Almö Bridge arch to a giant beer can, but then again it is inexcusable by the designer to *propose* and for the bridge owner/authorities to *accept* such a "collision-prone" structure, which, in addition, was so sensitive even to very small damages. The unstiffened cross-section (beside the ring stiffeners at the load-application points) was for sure too optimal for its own good. Would longitudinal stiffeners have been applied inside the tube then the load-carrying capacity would not only have increased but, most importantly, also the resistance to horizontal blows as well (Fig. 19.16).

Back in the 1800's longitudinal stiffeners (together with ring stiffeners) were always used in the design of tubular bridge members, as was the case for the Forth Bridge (see Fig. 3.23). But then, back in those days, it was natural to use longitudinal stiffeners as the tubes were built up of riveted curved plate parts which had to be joined together longitudinally.

Buckling of the shell in between the longitudinal stiffeners – if such are applied – no longer results in the collapse of the entire cross-section as the damaged zone now is "confined". As a matter of fact, there is also a capacity for post-critical reserve effects to develop as the edge zones close to the longitudinal stiffeners are able to carry load above the critical buckling load. Would the Almö Bridge tubes have been stiffened in the longitudinal direction then perhaps the bridge would have survived the collapse (just as the case would have been for a truss arch bridge, which normally consist of

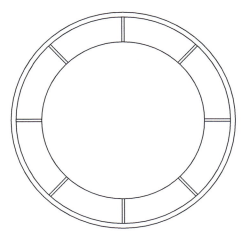

Fig. 19.16 Longitudinal stiffeners in between the ring stiffeners would markedly have increased the stability of the membrane.

Fig. 19.17 The new bridge between Almön and Källön, named the (New) Tjörn Bridge.

elements not sensitive to local impact damages). But still, both these alternatives (a longitudinally stiffened tubular arch bridge and a truss bridge) are not a good choice with respect to bridge collisions where major impact loading is a probability, as the case was here (with the restricted sailing width). But, as has been stated before, it did not matter whether the blow from the gantry crane was big or small; the unstiffened tubes were not strong enough to resist either case.

To accommodate for the transports to and fro the island of Tjörn a ferry was soon taken into service, and the planning of a new bridge was immediately started. Due to the intact approach ramps (see Fig. 19.8) – which could be used during construction (but not being part in the final bridge though) – and the fact that the project was given absolute priority from the authorities, the construction of the new bridge could begin already in August the same year. This time the choice fell upon a cable-stayed bridge with a main span of 366 metres, giving no restrictions whatsoever to the sailing width (Fig. 19.17).

This new bridge was completed and taken into service on 9 November 1981, only some fifteen months after the construction began, and less than two years after the failure of the old Almö Bridge (the fastest ever construction in the world of a major bridge project?). As extra safety precautions the pylons were positioned more than 40 metres up on the shore (hence the longer span length) and the free height was increased 3.5 metres. By choosing a suspended structure – which was the inevitable solution given the background – a bridge type that back in the 1950's had been considered, but rejected because of the pylons, was finally being built.

Despite the increase in free sailing height, the Tjörn Bridge was hit by a ship in 1983 in approximately the same manner as the old bridge. It was a cargo ship of nearly the same size as the Star Clipper that accidentally had forgotten to lower its loading crane before passing under the bridge. But this time the superstructure was able to withstand the impact loading with only minor damages, being robust as it was to this kind of loading, which the old bridge for sure was not.

Chapter 20

Sgt. Aubrey Cosens VC Memorial Bridge

In the small Canadian town of Latchford, in the province of Ontario (Fig. 20.1), a 106.7 metre long compression arch suspended-deck bridge spanning the Montreal River was built in October 1960 (Figs. 20.2 and 20.3).

In 1986 the bridge was given its present name in order to honour the World War II soldier Sergeant Aubrey Cosens from Latchford, who received the Victoria Cross posthumously for bravery in a battle with the Germans, February 1945.

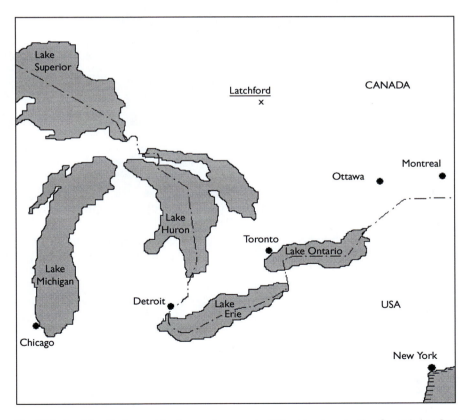

Fig. 20.1 Latchford is located in the southern part of Canada, close to the Great Lakes Region.

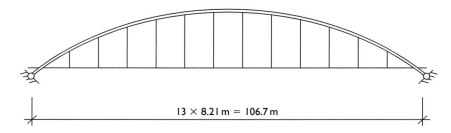

Fig. 20.2 The single-span bridge is 106.7 metres long.

Fig. 20.3 The Sgt. Aubrey Cosens VC Memorial Bridge. (www.thekingshighway.ca/latchford.html, with kind permission of Bala Tharmabala, Ministry of Transportation, Ontario)

The bridge deck was supported by a stringer-to-floor-beam system (six stringers in each bay) (Fig. 20.4), where the floor-beams (twelve in all) were supported by vertical hangers which were suspended from the arch (Fig. 20.5).

In order to avoid bending each hanger was made free to rotate about a transverse axis in two locations, down below close to the floor-beam and up above close to the arch (Figs. 20.5 and 20.6).

The last section of each hanger – close to the arch – was made up of a steel rod which was attached to the arch (inside the profile) by a connection detail (positioned at a diaphragm plate), where the rod was safely anchored by two nuts and a washer on the rod's threaded end (Fig. 20.7).

There are two local actions to consider when the hangers are forced to rotate (besides global bending deformation of the arch). First there is the effect coming from unsymmetrical loading taking the flexibility of the *stringers* into account (Fig. 20.8), and secondly, there is the effect from the flexibility of the *hanger* itself (Fig. 20.9).

Back in 1960, when the bridge was constructed, the use of two hinges ensured that no bending whatsoever was induced into the hangers, and whatever the rotation the

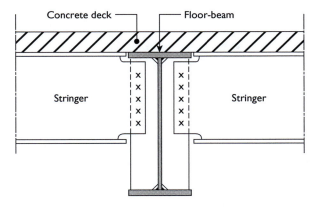

Fig. 20.4 Stringer-to-floor-beam connection (principal layout).

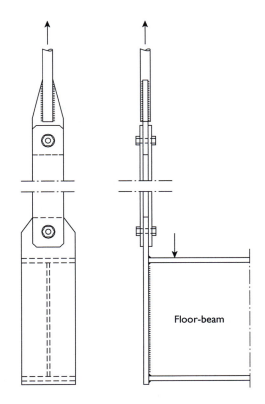

Fig. 20.5 Hanger-to-floor-beam connection.

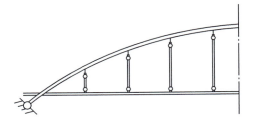

Fig. 20.6 Each vertical hanger was pin-jointed in two locations.

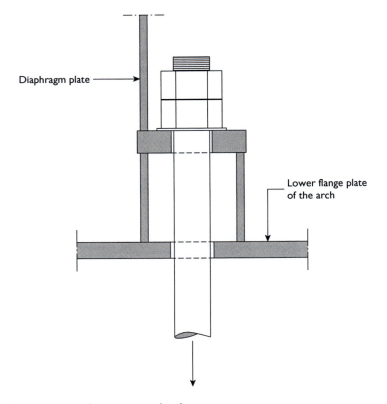

Fig. 20.7 Hanger-to-arch connection detail.

upper rod was always kept straight. It must be assumed that for many years this vital function of the hangers – i.e. to be able to rotate without producing any bending – was kept intact, however, as time went on there was a gradual deterioration due to lack of inspection and maintenance.

The following description – which is made from the perspective of the bridge – is a reconstruction of the course of events leading up to the partial collapse that took place in 2003. The step-by-step breakdown process has been verified in the post-failure

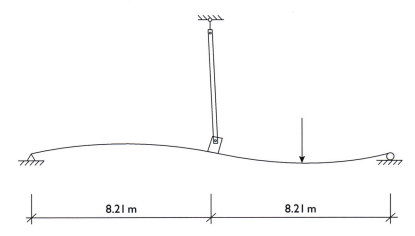

Fig. 20.8 Rotation of the hanger due to unsymmetrical loading.

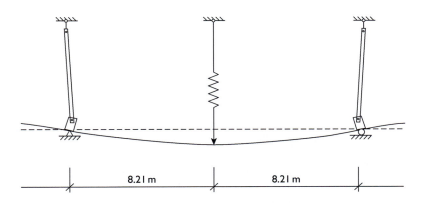

Fig. 20.9 When a floor-beam is subjected to a concentrated load, the difference in deflection makes the adjacent hangers rotate. For the sake of simplicity the adjacent hangers have been kept rigid in the description above, which, of course, they are not.

analysis that was made, but during the period prior to the failure no indications of any problems or the hangers malfunctioning were observed by inspection personnel.

The bridge served its purpose well for many years, and perhaps this led to the belief that the structure more or less could be left unattended. In particular the hangers, with their special design of having two hinges, were taken for granted to be functioning – without any extra maintenance procedures needed – as they were supposed to, i.e. to rotate without producing any bending. Also, as there were only very small rotations, hardly visible for the naked eye, it was very difficult to notice any gradual change in the capability of the hangers to accommodate for the necessary rotation capacity. So when the pin joints of the hangers gradually rusted stuck no inspection personnel was able to discover this lack of function – perhaps this absolutely vital function was not fully understood as well. The function of the hangers was just to transfer the vertical load

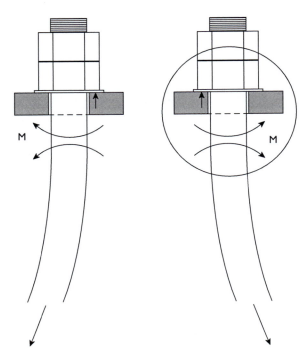

Fig. 20.10 Secondary bending of the hangers due to the inability of the pin joints to rotate.

to the arch, and as the hangers always were "vertical" there could not be any problem, could it? But, the inability of the pin joints to adjust to the rolling load on the bridge deck, led to back-and-forth bending deformations of the hangers (Fig. 20.10).

Would the connection to the arch have been a true hinge then there would not have been any problem, however, as the reaction force became eccentrically applied – the contact pressure of the washer moving from side to side – this connection acted more or less like a fixed end. The stiffness of the hanger itself is very small, but close to the connection to the arch the rigidity increases dramatically (because of the "fixed-end condition"). Possibly there was also additional secondary bending produced in the transverse direction because of bending deformations of the floor-beam, but most probably these stresses were negligible in comparison to the longitudinal effect.

It is true that these secondary bending deformations (in the longitudinal direction) were very small, but over the long run (read: after a sufficient number of loading cycles, perhaps several millions) a fatigue crack was initiated in one of the hangers (Fig. 20.11).

If such a crack is left unattended (and it is impossible in this case to do anything else, as this connection is "enclosed" and hidden inside the arch) then it will propagate until the critical crack length is reached and the net-section will fracture in a brittle manner. And it would not have mattered if the steel would have been more fracture tough (read: ductile, being of a more modern standard) it would still have failed, the process taking just a little longer time.

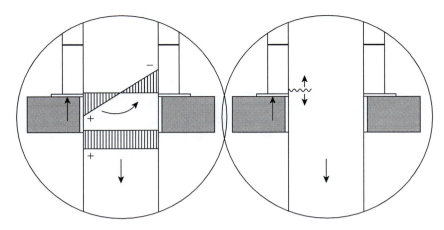

Fig. 20.11 Secondary bending stresses, in addition to the normal tension stress, will eventually initiate a fatigue crack (see also Fig. 20.10).

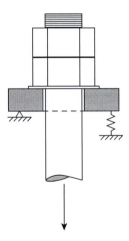

Fig. 20.12 The support conditions – a difference in stiffness due to the single diaphragm plate (see also Fig. 20.7).

It was the hanger closest to the abutment of the northwest corner of the bridge that failed first, being the shortest (i.e. more axial stiff then the rest of the hangers), thus experiencing more dynamic effects than the longer and softer hangers. The rolling load coming from north (or south for that matter) will also produce a more type of "shock loading" to the first hanger than the ones more centrally located in the bridge where the load is more evenly distributed and transferred in a more gradual manner. In addition to these dynamic effects, there was also a small defect present in the threaded part of the rod where the crack had originated from. Most likely the difference in stiffness of the support plate "bearings" also contributed to the initiation of the crack, probably then starting from the stiffer side at the diaphragm plate (Fig. 20.12).

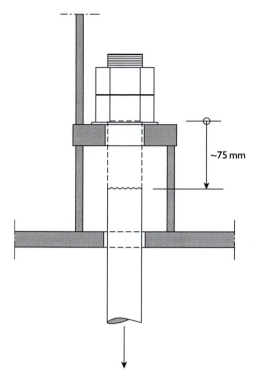

Fig. 20.13 Fracture of Hanger 1 (the one closest to the abutment on the northwest corner of the bridge). See also figure 20.7 as a comparison.

During the early part of the 1990's the crack in the upper threaded end of the Hanger 1 rod (the one closest to the abutment at the northwest corner of the bridge) continued to propagate and increased in length for each new loading cycle from heavy traffic, until it fractured completely around 1996–1997 when the net-section of the rod was about 20% of the original cross-sectional area. The hanger immediately lost its load-carrying capacity, and it fell down, but was stopped after some 75 millimetres (Fig. 20.13).

The hanger was stopped from falling down completely (i.e. being pulled through the hole in the flange plate) by the bending stiffness of the stringer-to-floor-beam system, and possibly also to some extent by the flange plate of the arch as the upper end of the rod was slightly enlarged in comparison to the diameter below the connection (most certainly a decision by the designers to compensate for the reduction in cross-sectional area because of the threading – not shown in the figure above though). The hanger was still standing "intact" (to the observer), but was nothing more than a "flagpole" (free at the top, and connected only at the bottom to the floor-beam) without any function whatsoever (it was there, but at the same time it was not). As Hanger 1 was still standing in its vertical position the fracture was not discovered – the only thing (at deck level) that could indicate the failure was the deflection of the deck itself, but it was too small to be noticed. Not even the painting crew, which repainted the bridge in

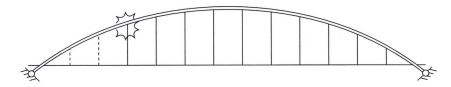

Fig. 20.14   Hanger 3 broke during the passage of a tractor-trailer on 14 January 2003.

1998, reacted upon the strange fact that the Hanger 1 rod was 75 millimetres longer than the rest of the hanger rods.

As the load-carrying capacity of Hanger 1 now was zero the weight of the deck and the load from the traffic had to be transferred to the adjacent hanger (Hanger 2). Very soon a fatigue crack was initiated in Hanger 2 as well (if not being present already prior to the fracture of Hanger 1) and this hanger fractured around 2000–2001. Now there was a situation where two hangers were completely gone (Hangers 1 and 2), and where Hanger 3 had become heavily strained. The increased deflection of the deck was still not noticed, but hidden below the concrete deck extensive fatigue cracking could have been detected (if an inspection had been carried out that is) in the stringer-to-floor-beam connection at the Hanger 1 position due to the excessive flexural bending of the stringers. But still the concrete deck and the stringer-to-floor-beam system were able to bridge over this local (unintended) gap, like a bridge within the bridge. And not to forget, the hangers on the north*east* corner (i.e. on the other side of the bridge deck) was still functioning as they should, undamaged as they were. The bridge withstood the fracture of two hangers without failing completely, thus showing a high degree of structural redundancy, which is a very important and needed characteristic of a bridge – it is absolutely vital that the bridge is given such a design, if possible, that it responds in a robust way to damages.

During the passage of a southbound tractor-trailer on 14 January 2003, at around 3 in the afternoon, Hanger 3 finally broke (Fig. 20.14).

An extremely low temperature at the time of the trailer passage ($-25°$C) contributed to the brittle fracture of Hanger 3, but it would have failed sooner or later anyhow (a fatigue crack had been initiated in this hanger as well).

When Hanger 3 fractured the deck collapsed completely and fell down some two metres (the driver of the trailer managed to get safely off the bridge though, and no one was injured) (Fig. 20.15).

The visible sign that the pin-joints of hangers had rusted stuck, was that the hangers were still standing up after the collapse, instead of being folded together (Fig. 20.16, see also figures 20.5 and 20.6).

A similar incident occurred in Sweden in the 1980's, when a 61 metre long tied-arch railway bridge in Kusfors over the Skellefte Älv – some 300 kilometres to the north of Sandö Bridge, see figure 6.1 – experienced fatigue cracking in the vertical hangers. During a routine inspection the personnel noticed a "rattling sound" when a passenger train was passing the bridge, and to their great surprise they discovered that one of the hangers was completely broken at the top, but still standing in its vertical position (as it still was supported laterally, like the broken hangers of the Aubrey

Fig. 20.15 The (partial) collapse of the Sgt. Aubrey Cosens VC Memorial Bridge. (http:speaker. northernontario.ca/content/Feature%20Story/C%20Front%2011:01:06.pdf)

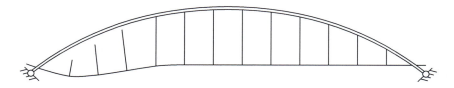

Fig. 20.16 The three fractured hangers were still standing up after the collapse (see also Fig. 20.15).

Cosens Bridge). The inspection personnel knew that the heaviest goods train on the line ("Stålpilen" – the Steel Arrow) soon was to pass, and they did not know what to do. Was the bridge able to carry the load of this very heavy train (with an axle weight of up to 22.5 tons)? As the bridge had carried the passenger train (and perhaps additional trains before this inspection), it was decided that the goods train was allowed to pass as well, but at a considerably reduced speed. Everything went well, and the bridge was later repaired where all the hangers were replaced with new ones. However, in contrast to the Aubrey Cosens Bridge, where the bridge still carried traffic load even though two hangers were broke, the Kusfors Bridge most probably could not have carried

Fig. 20.17 The rehabilitated Aubrey Cosens Bridge at Latchford, September 2007 (with kind permission of Cameron Bevers, photographer and copyright holder).

railway traffic (especially heavy goods trains) with more than one hanger broke. In this case it was very lucky indeed that the hanger failure was discovered in time.

The Aubrey Cosens Bridge was eventually repaired and again taken into service in 2005, this time using four cable wires in each hanger location instead of a bending stiff rod (Fig. 20.17). The use of several parallel cables (instead of a single thick one) is also a good choice with respect to safety and robustness – the loss of one wire would not mean the entire failure of the hanger as the remaining wires would take over the load.

As an alternative to the connection detail at the top to the arch in the *old* bridge – as it was originally designed – the use of a *rocker bearing* (a true one) would most probably have saved not only the Aubrey Cosens Bridge, but also the Kusfors Bridge. However, if the pin joints of the hangers would have been properly inspected and maintained then a rocker bearing solution is superfluous. But as a safeguard to any hidden friction problems, the upper pin joint – of the original design – could be omitted and a rocker bearing at the top be used instead, as a safe life solution (Fig. 20.18).

In retrospect one could say that the failure of the Aubrey Cosens Bridge was because of the same secondary constraints that he designers back in 1960 very hard tried to avoid – shaping the hangers with two pin joints each – but their effort was to no good as the function of the hangers, because of lack of inspection and maintenance, gradually deteriorated.

And finally, we see the same life-span expectancy of the Aubrey Cosens Bridge and the Kusfors Bridge, as we did for the Point Pleasant Bridge and the Reichsbrücke (all containing eye-bars) – being left unattended with respect to proper maintenance that

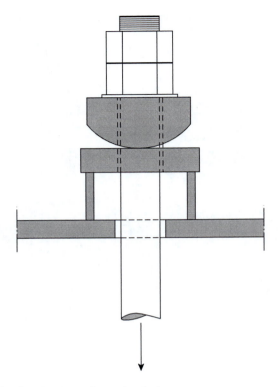

Fig. 20.18 A rocker bearing as an alternative design.

is. Both the Aubrey Cosens Bridge and the Kusfors Bridge failed after approximately 40 years (the Kusfors Bridge was built in 1943), as the latter two bridges also did. With respect to failure cause we could leave out the Reichsbrücke in our comparison, but the other three failed in fatigue due to pin joints not functioning as they should.

Literature

Dee Bridge

[1] Steinman, D.B. – Watson, S.R.: Bridges and their builders. Dover Publications Inc., New York 1957.

[2] Cornell, E.: Byggnadstekniken – metoder och idéer genom tiderna. Bygg-förbundet 1970.

[3] Hopkins, H.J.: A span of bridges – an illustrated history. Praeger Publishers, New York 1970.

[4] Stålbygge. Utdrag ur Stålbyggnad, programskrift 11 från Statens råd för byggnadsforskning, Svenska Reproduktions AB, Stockholm 1971.

[5] Wallin, L.: Att bygga i stål – moderna teknik med gamla anor. Särtryck ur Daedalus – Tekniska Muséets årsbok 1973, Stålbyggnadsinstitutet, Publ. 43, Stockholm 1973.

[6] Berridge, P.S.A.: The girder bridge after Brunel and others. Robert Maxwell Publisher, London 1969.

[7] Beckett, D.: Stephenson's Britain. David & Charles, London 1984.

[8] Petroski, H.: Success syndrome: The collapse of the Dee Bridge. Civil Engineering, April 1994.

[9] Lewis, P.R. – Gagg, C.: Aesthetics versus function: The fall of the Dee Bridge, 1847. Interdisciplinary Science Reviews, Vol. 29, No. 2, 2004.

[10] Petroski, H.: Past and present failures. American Scientist, Vol. 92, Nov/Dec 2004.

Ashtabula Bridge

[11] On the failure of the Ashtabula Bridge. Transactions of the American Society of Civil Engineers (ASCE), 1877; Discussions on subjects presented at the ninth annual convention.

[12] Gasparini, D.A. – Fields, M.: Collapse of Ashtabula Bridge on December 29, 1876. Journal of Performance of Constructed Facilities (ASCE), Vol. 7, No. 2, May 1993.

[13] The Ashtabula Train Disaster: http://www.deadohio.com/AshTrain.htm

[14] The Ashtabula Bridge Disaster: http://home.alltel.net/arhf/bridge.htm

[15] Ashtabula River Railroad Disaster: http://www.reference.com/browse/wiki/Ashtabula_River_Railroad_Disaster

[16] Ashtabula Bridge Disaster: http://www.prairieghosts.com/rr_disaster.html

[17]   The Ashtabula Disaster. Harper's Weekly – January 20, 1877. http://www.catskillarchive.com/rrextra/wkasht.html

[18]   Notes on Railroad Accidents: http://www.catskillarchive.com/rrextra/wkbkch11.html

[19]   Ashtabula Bridge Disaster, Philip P. Bliss – Christian Biography Resources: http://www.wholesomewords.org/biography/biobliss5.html

[20]   Petroski, H.: Engineers of Dreams – Great Bridge Builders and the Spanning of America. Alfred A. Knopf, New York 1995, pp 96–97.

Tay Bridge

[21]   Hammond, R.: The Forth Bridge and its builders. Eyre & Spottiswoode (Publishers) Ltd, London 1964.

[22]   Lewis, P.R. – Reynolds, K.: Forensic engineering: a reappraisal of the Tay Bridge disaster. Interdisciplinary Science Reviews, Vol. 27, No. 4, 2002.

[23]   Pinsdorf, M.K.: Engineering Dreams Into Disaster: History of the Tay Bridge. Business and Economic History, Vol. 26, No. 2, 1997.

[24]   Berridge, P.S.A.: The girder bridge after Brunel and others. Robert Maxwell Publisher, London 1969.

[25]   Pugsley, A.: The Safety of Structures. Edward Arnold (Publishers) Ltd, London 1966.

[26]   Hammond, R.: Engineering Structural Failure – The causes and results of failure in modern structures of various types. Odhams Press Limited, London 1956.

[27]   Jones, D.R.H.: The Tay Bridge. Engineering Materials 3: Materials Failure Analysis: Case Studies and Design Implications. Chapter 27 and 28. Pergamon Press 1993.

[28]   Tom Martin's Tay Bridge Disaster Web Pages. http://taybridgedisaster.co.uk

[29]   Forensic Engineering: The Tay Bridge Disaster. www.open2.net/forensic_engineering/riddle/riddle_01.htm

[30]   Martin, T. – Macleod, I.A.: The Tay rail bridge disaster – a reappraisal based on modern analysis methods. Proc. Instn. Civ. Engrs., Civil Engineering, 108, May 1995.

Quebec Bridge

[31]   Hammond, R.: Engineering Structural Failure – The causes and results of failure in modern structures of various types. Odhams Press Limited, London 1956.

[32]   Pugsley, A.: The Safety of Structures. Edward Arnold (Publishers) Ltd, London 1966.

[33]   Steinman, D.B. – Watson, S.R.: Bridges and their builders. Dover Publications Inc., New York 1957.

[34]   Hammond, R.: The Forth Bridge and its builders. Eyre & Spottiswoode (Publishers) Ltd, London 1964.

[35]   Hopkins, H.J.: A span of bridges – an illustrated history. Praeger Publishers, New York 1970.

[36]   The Quebec bridge. http://geocities.com/Colosseum/Bench/918/cssq/pont.html

[37]   Quebec Bridge. http://en.wikipedia.org/wiki/Quebec_Bridge

[38]   Triumphs of Engineering. Odhams Press Limited, London 1946, pp 140–145.

[39]   Top 125 years in ENR history. http://enr.construction.com/advertise/aboutUs/125enrHistory/990308.asp

[40]   Världens största cantilever-bro. Teknisk Tidskrift, 2 Sept. 1916, pp 332–333.

[41]   Bergfelt, A. – Edlund, B.: Stål- och träbyggnad. Del 1 Bärande konstruktioner. Chalmers tekniska högskola, Institutionen för konstruktionsteknik, Avd. för stål- och träbyggnad, Publ. S69:1, Göteborg 1969, pp 40–41.

[42]   Den nya katastrofen vid Quebec-bron. Teknisk Tidskrift, 4 Nov. 1916, pp 408–411.

[43]   Petroski, H.: Engineers of Dreams – Great Bridge Builders and the Spanning of America. Alfred A. Knopf, New York 1995, p 97.

[44]   Roddis, W.M.: Structural Failures and Engineering Ethics. Journal of Structural Engineering, Vol. 119, No. 5, May, 1993, pp 1539–1546.

[45]   The First Quebec Bridge Disaster. http://www.civing.carleton.ca/ECL/reports/ECL270/Disaster.html

[46]   When Haste Makes Waste – The Quebec Bridge. By Johnny Beauvais. http://www.tbirdonline.com/quebec1.htm

[47]   Francis, A.J.: Introducing Structures – civil and structural engineering, building and architecture. Ellis Horwood Series in Civil Engineering, 1989.

[48]   Pearson, C. – Delatte, N.: Collapse of the Quebec Bridge, 1907. Journal of Performance of Constructed Facilities (ASCE), February 2006, pp 84–91.

Hasselt Bridge

[49]   Stålbygge. Utdrag ur Stålbyggnad, programskrift 11 från Statens råd för byggnadsforskning, Svenska Reproduktions AB, Stockholm 1971.

[50]   Shafie, M.A. – Sabardin, L.N.: Ageing and brittle fracture in steel. Chalmers University of Technology, Department of Structural Engineering, Division of Steel and Timber Structures, Int. skr. S 97:8, Göteborg 1997.

[51]   Bridge Failure Database. http://www.bridgeforum.org/dir/collapse/bridge

[52]   Smith, D.W.: Bridge failures. Proc. Instn Civ. Engrs, Part 1, 1976, 60, Aug., pp 367–382.

[53]   Johansson, G.: Stålbyggnad V3. Chalmers tekniska högskola, Institutionen för konstruktionsteknik, Avd. för stål- och träbyggnad, Kompendium 02:3, Göteborg 2002.

[54]   Voormann, F. – Pfeifer, M. – Trautz, M.: Die ersten geschweissten Stahlbrücken in Deutschland – Über die wechselvollen Anfänge der Schweisstechnik. Stahlbau 75 (2006), Heft 4, pp 287–297.

[55]   Collapse of the Hasselt Bridge. Civil Engineering (London), Vol. 34, No. 393, 1939.

[56]   Hammond, R.: Engineering Structural Failure – The causes and results of failure in modern structures of various types. Odhams Press Limited, London 1956.

[57]   McGuire, W.: Steel Structures. Prentice-Hall, Inc. / Englewood Cliffs, N.J., 1968.

[58]   Åkesson, B.: Fatigue Life in Riveted Railway Bridges. Chalmers University of Technology, Department of Structural Engineering, Publ. S 94:6, Göteborg, 1994.

Sandö Bridge

[59] Granholm, H.: Sandöbrons bågställning". Chalmers tekniska högskolas handlingar, Avd. Väg- och Vattenbyggnad, Byggnadsteknik, Nr. 239, Scandinavian University Books, Göteborg, 1961.

[60] Cornell, E.: Byggnadstekniken – metoder och idéer genom tiderna. Byggförbundet 1970.

[61] Sandöbron. A newspaper article in Expressen, August 26, 1979.

[62] Bygden vid Sandöbron. http://www.kramforsbygder.com/bygdenvid-sandobron

[63] Ådalens hjärta. http://webtelia.com/~u60105008/adalen_1.html

Tacoma Narrows Bridge

[64] Selberg, A.: Svingninger i hengebruer. Teknisk Tidskrift, 23 Nov. 1946, pp 1201–1207.

[65] Pugsley, A.: The Safety of Structures. Edward Arnold (Publishers) Ltd, London 1966.

[66] Asplund, S.O.: Deflection Theory Analysis of Suspension Bridges. International Association for Bridge and Structural Engineering (IABSE), Zürich, 1949.

[67] McGuire, W.: Steel Structures. Prentice-Hall, Inc. / Englewood Cliffs, N.J., 1968.

[68] Hammond, R.: Engineering Structural Failure – The causes and results of failure in modern structures of various types. Odhams Press Limited, London 1956.

[69] Hult, J.: Laster och brott. Almqvist & Wiksell Förlag AB, Stockholm, 1969.

[70] Stålbyggnad – Utveckling och forskningsbehov. Programskrift 11, Statens råd för byggnadsforskning, Svenska Reproduktions AB, Stockholm, 1970.

[71] Steinman, D.B. – Watson, S.R.: Bridges and their builders. Dover Publications Inc., New York 1957.

[72] Nilsson, S.: Om svagt avstyvade hängbroar och deras beräkning genom modellförsök. Teknisk Tidskrift, 23 Feb. 1935, pp 17–20.

[73] Nakao, M.: Collapse of Tacoma Narrows Bridge – November 7, 1940 in Tacoma, Washington, USA. Failure Knowledge Database: http://shippai.jst.go.jp/en/Search

[74] Pugsley, A.: The Theory of Suspension Bridges. Edward Arnold (publishers) LTD, Great Britain, 1968.

[75] Lorentsen, M. – Sundquist, H.: Hängkonstruktioner. Kompendium i Brobyggnad. Institutionen för Byggkonstruktion, Kungliga Tekniska Högskolan, Rapport 18, Stockholm, 1995.

[76] Sibly, P.G. – Walker, A.C.: Structural accidents and their causes. Proc. Instn. Civ. Engrs., Part 1, 1977, 62, May, pp 191–208.

[77] ESDEP Lectures. European Steel Design Education Program. WG 15B Structural Systems: Bridges, Lecture 15B.9: Suspension bridges. The Steel Construction Institute, Berkshire, England.

Peace River Bridge

[78] Jean, W. D.: A historical perspective, documentation, and input on Peace River Bridge. Term paper in a Foundation course at the University of Tennessee at Chattanooga, 2006.

[79] ESDEP Lectures. European Steel Design Education Program. WG 15B Structural Systems: Bridges, Lecture 15B.9: Suspension bridges. The Steel Construction Institute, Berkshire, England.

[80] Story of the Alaska Highway. http://www.ourpage.net/341st/AH/

[81] The Peace River bridge. http://www.ourpage.net/341st/Bridge/

[82] Anchorage Slip Wrecks Suspension Bridge. Engineering News Record, October 24, 1957, p 26.

Second Narrows Bridge

[83] McGuire, W.: Steel Structures. Prentice-Hall, Inc./Englewood Cliffs, N.J., 1968.

[84] Åkesson, B.: Plate Buckling in Bridges and other Structures. Taylor & Francis/Balkema, Leiden, 2007.

[85] Ironworkers Memorial Second Narrows Crossing. http://en.structurae.de/structures/data/index.cfm?id=s0007234

[86] Ironworkers Memorial Second Narrows Crossing. http://en.wikipedia.org/wiki/Ironworkers_Memorial_Second_Narrows_Crossing

[87] Second Narrows Bridge. http://en.wikipedia.org/wiki/Second_Narrows_Bridge

[88] The History of Metropolitan Vancouver – Collapse of the Second Narrows Bridge. http://www.vancouverhistory.ca/archives_second_narrows.htm

[89] Burrard Inlet. http://en.wikipedia.org/wiki/Burrard_Inlet

Kings Bridge

[90] Report of Royal Commission into the failure of the Kings Bridge. Victoria, 1963.

[91] Jones, D.R.H.: Fast fracture of a motorway bridge. Engineering Materials 3: Materials Failure Analysis: Case Studies and Design Implications. Chapter 15. Pergamon Press, 1993.

[92] Ekström, H. – Lantz, L-E.: Sprödbrottet i Kings Bridge, Melbourne. Seminarieuppsats i Byggnadsteknik II, Fortsättningskurs Ht 1972, Chalmers tekniska högskola, Institutionen för konstruktionsteknik, Stål- och träbyggnad. Göteborg, 1972.

[93] Alpsten, G. – Ingwall, C.T.: Förstärkningsplåtar försvagningsplåtar? Ett inlägg i diskussionen om transversella svetsar i dragzonen. Väg- och vattenbyggaren 9, 1971.

[94] ESDEP Lectures. European Steel Design Education Program. WG 1B, Lecture 1B.8: Learning from failures. The Steel Construction Institute, Berkshire, England.

[95] Stålbyggnad – Utveckling och forskningsbehov. Programskrift 11, Statens råd för byggnadsforskning, Svenska Reproduktions AB, Stockholm, 1970.

Point Pleasant Bridge

[96] Dicker, D.: Point Pleasant Bridge collapse mechanism analyzed. Civil Engineering (ASCE), July 1971.

[97] Cause of Silver Bridge collapse studied. Civil Engineering (ASCE), December 1968.

[98] Scheffey, C.F.: Pt. Pleasant Bridge collapse – conclusion of the federal study. Civil Engineering (ASCE), July 1971.

[99] Lichtenstein, A.G.: The Silver Bridge Collapse Recounted. Journal of Performance of Constructed Facilities, Vol. 7, No. 4, November 1993.

[100] LeRose, C.: The Collapse of the Silver Bridge. West Virginia Historical Society Quarterly, Volume XV, No. 4, October 2001.

[101] Silver bridge. http://en.wikipedia.org/wiki/Silver_Bridge

[102] David B. Steinman. http://en.wikipedia.org/wiki/David_Barnard_Steinman

[103] Hercilio Luz Bridge. http://en.wikipedia.org/wiki/Hercilio_Luz_Bridge

[104] Silver Memorial Bridge. http://wikipedia.org/wiki/Silver_Memorial_Bridge

Fourth Danube Bridge

[105] Heckel, R.: The Fourth Danube Bridge in Vienna – Damage and Repair. Conference Proceedings, "Developments in Bridge Design and Construction", University College Cardiff, 1971.

[106] ESDEP Lectures. European Steel Design Education Program. WG 1B, Lecture 1B.8: Learning from failures. The Steel Construction Institute, Berkshire, England.

[107] Bergfelt, A.: Undersökning av bärförmågan hos lådbalkar av slanka plåtar, speciellt med avseende på instabilitetsfenomen. Chalmers University of Technology, Department of Structural Engineering, Division of Steel and Timber Structures, Int. skr. S 78:7, Göteborg, Sweden.

[108] Aurell, T. – Englöv, P.: Olyckor vid montering av stora lådbroar. Väg- och vattenbyggaren 5, 1973.

[109] Roik, K.: Betrachtungen über die Bruchursachen der neuen Wiener Donaubrücke. Tiefbau, 12, Dezember, 1970.

[110] Herzog, M.A.M.: Simplified Design of Unstiffened and Stiffened Plates. Journal of Structural Engineering, Vol. 113, No. 10, October, 1987.

[111] Åkesson, B.: Plate Buckling in Bridges and other Structures. Taylor & Francis/ Balkema, Leiden, 2007.

Britannia Bridge

[112] Ryall, J.M.: Britannia Bridge, North Wales: Concept, Analysis, Design and Construction. Proceedings of the International Historic Bridge Conference, Columbus, Ohio, USA, August 27–29, 1992.

[113] Hopkins, H.J.: A Span of Bridges – an illustrated history. Praeger Publishers, Inc., New York, 1970.

[114] Werner, E.: Die Britannia- und Conway-Röhren-brücke. Werner-Verlag, Düsseldorf, 1969.

[115] Cornell, E.: Byggnadstekniken – Metoder och idéer genom tiderna. Byggförbundet, 1970.

[116] Steinman, D.B. – Watson, S.R.: Bridges and their builders. Dover Publications Inc., New York 1957.

[117] de Maré. E.: Your Book of Bridges. Faber and Faber, London, 1963.

[118] Beckett, D.: Stephenson's Britain. David & Charles, London, 1984.

[119] Sandström, G.E.: Byggarna – Teknik och kultur från vasshus och pyramid till järnväg och högdamm. Bokförlaget Forum, Stockholm, 1968.

[120] Bailey, M.R.: Robert Stephenson. Steel Construction Today, September 1991.

[121] Cornwell, E.L.: Britannia Bridge reopening restores Holyhead route. Modern Railways, February 1972.

[122] Husband, R.W.: The Britannia Rail and Road Bridge in North Wales (United Kingdom). IABSE Structures C-16/81, IABSE Periodica 1/1981.

[123] Timoshenko, S.P.: History of Strength of Materials. McGraw-Hill, New York, 1953.

[124] Britannia Bridge Official Fire Report. http://www.2d53.co.uk/britanniabridge/Fire%20Report.htm

Cleddau Bridge

[125] ESDEP Lectures. European Steel Design Education Program. WG 1B, Lecture 1B.8: Learning from failures. The Steel Construction Institute, Berkshire, England.

[126] Bergfelt, A.: Undersökning av bärförmågan hos lådbalkar av slanka plåtar, speciellt med avseende på instabilitetsfenomen. Chalmers University of Technology, Department of Structural Engineering, Division of Steel and Timber Structures, Int. skr. S 78:7, Göteborg, Sweden.

[127] Aurell, T. – Englöv, P.: Olyckor vid montering av stora lådbroar. Väg- och vattenbyggaren 5, 1973.

[128] Cantilevered box girder bridge collapses during construction. Engineering News Record (ENR), June 11, 1970.

[129] Åkesson, B.: Plate Buckling in Bridges and other Structures. Taylor & Francis/Balkema, Leiden, 2007.

West Gate Bridge

[130] Report of Royal Commission into the Failure of West Gate Bridge. The West Gate Bridge Royal Commission Act 1970, No. 7989, Victoria, 1971. VPRS No. 2591/PO, Unit 14, Transport of Proceedings.

[131] ESDEP Lectures. European Steel Design Education Program. WG 1B, Lecture 1B.8: Learning from failures. The Steel Construction Institute, Berkshire, England.

[132] Bergfelt, A.: Undersökning av bärförmågan hos lådbalkar av slanka plåtar, speciellt med avseende på instabilitetsfenomen. Chalmers University of Technology, Department of Structural Engineering, Division of Steel and Timber Structures, Int. skr. S 78:7, Göteborg, Sweden.

[133]   Aurell, T. – Englöv, P.: Olyckor vid montering av stora lådbroar. Väg- och vattenbyggaren 5, 1973.

[134]   Edlund, B.: Katastrofen vid West Gate Bridge, Melbourne – ett dystert treårsminne. Väg- och vattenbyggaren 5, 1973.

[135]   Bignell, V.: Catastrophic Failures. Case Study 5: The West Gate Bridge Collapse. The Open University Press, New York.

[136]   Paul, A.M.: När den stora bron rasade. Det Bästa, October 1972. Summary (in Swedish) from Sunday Independent, 18 June 1972.

[137]   Åkesson, B.: Plate Buckling in Bridges and other Structures. Taylor & Francis/ Balkema, Leiden, 2007.

[138]   Disaster at West Gate.   http://www.prov.vic.gov.au/exhibs/westgate/ welcome.htm

[139]   West Gate Bridge. http://en.wikipedia.org/wiki/West_Gate_Bridge

Rhine Bridge

[140]   Maquoi, R. – Massonnet, C.: Annales de Travaux Publics de Belgique. Chapter 5. Le Pont sur le Rhin a Coblence. No. 2, 1972.

[141]   ESDEP Lectures. European Steel Design Education Program. WG 1B, Lecture 1B.8: Learning from failures. The Steel Construction Institute, Berkshire, England.

[142]   Bergfelt, A.: Undersökning av bärförmågan hos lådbalkar av slanka plåtar, speciellt med avseende på instabilitetsfenomen. Chalmers University of Technology, Department of Structural Engineering, Division of Steel and Timber Structures, Int. skr. S 78:7, Göteborg, Sweden.

[143]   Aurell, T. – Englöv, P.: Olyckor vid montering av stora lådbroar. Väg- och vattenbyggaren 5, 1973.

[144]   Åkesson, B.: Plate Buckling in Bridges and other Structures. Taylor & Francis/ Balkema, Leiden, 2007.

[145]   Eurocode 3: Design of steel structures – Part 1–1: General rules and rules for buildings. EN 1993-1-1: 2005.

[146]   Eurocode 3: Design of steel structures – Part 1–5: Plated structural elements. EN 1993-1-5: 2006.

Zeulenroda Bridge

[147]   Ekardt, H-P.: Die Stauseebrücke Zeulenroda. Ein Schadensfall und seine Lehren für die Idee der Ingenieurverantwortung. Stahlbau 67 (1998), Heft 9.

[148]   Åkesson, B.: Plate Buckling in Bridges and other Structures. Taylor & Francis/ Balkema, Leiden, 2007.

[149]   Eurocode 3: Design of steel structures – Part 1–1: General rules and rules for buildings. EN 1993-1-1: 2005.

[150]   Eurocode 3: Design of steel structures – Part 1–5: Plated structural elements. EN 1993-1-5: 2006.

Reichsbrücke

[151]   Reiffenstuhl, H.: Collapse of the Viennese Rechsbrücke: Causes and Lessons. IABSE Symposium, Washington, 1982, Final Report.
[152]   Permanent monitoring of Viennese Reichsbrücke. History of Reichsbrücke. http://www.reichsbruecke.net/geschichte_e.php
[153]   Reichsbrücke. http://en.wikipedia.org/wiki/Reichsbr%C3%BCcke
[154]   Der Reichsbrückeneinsturz. http://www.wien-vienna.at/geschichte.php? ID=737
[155]   Eingestürzte Reichsbrücke in Wien, 1.8.1976. http://aeiou.iicm.tugraz.at/ aeiou.film.o/o262a

Almö Bridge

[156]   Tjörnbron. Vägverket (Sune Brodin), Adlink, 1984.
[157]   Hult, J.: Spänning och brott. Forskningens frontlinjer, Almqvist & Wiksell International, 1990.
[158]   Hartwig, H-J. – Hafke, B.: Die Bogenbrücke über den Askeröfjord. Der Stahlbau, Heft 10, Berlin, Oktober 1961.
[159]   Swedish bridge is brought down. New Civil Engineer (Magazine of the Institution of Civil Engineers), 24 January, 1980.
[160]   Den nya Tjörnbron. Väg- och vattenbyggaren 11–12, 1980.
[161]   Åkesson, B.: Plate Buckling in Bridges and other Structures. Taylor & Francis/ Balkema, Leiden, 2007.

Aubrey Cosens Bridge

[162]   Sgt. Aubrey Cosens V.C. Memorial Bridge over the Montreal River at Latchford. Investigation of Failure: Final Report, Ministry of Transport, Ontario, December 1, 2003. http://www.mto.gov.on.ca/english/engineering/ cosens/
[163]   The Latchford Bridge Failure (2003). http://www.thekingshighway.ca/ latchford.html
[164]   Cold snap. Partial collapse of a bridge over the Montreal River at the beginning of the year closed a critical artery of the Trans-Canada Highway. Bridge Design & Engineering, August 27, 2003. http://www.bridgeweb.com/ news/fullstory.php/aid/343/Cold_snap.html
[165]   Sergeant Aubrey Cosens V.C. Memorial Bridge. http://en.structurae.de/ structures/data/index.cfm?id=s0018160
[166]   Sgt. Aubrey Cosens VC Memorial Bridge. http://en.wikipedia.org/wiki/Sgt._Aubrey_Cosens_VC_Memorial_Bridge
[167]   Åkesson, B.: Äldre järnvägsbroar i stål – bärförmåga, kondition och livslängd. Chalmers University of Technology, Department of Structural Engineering, Division of Steel and Timber Structures, Publ. S 91:2, Göteborg 1991.

Picture and photo references

Front cover Chalmers University of Technology, Department of Structural Engineering, Steel and Timber Structures, Göteborg (slide picture of the Rhine Bridge in Koblenz).

Fig. 1.1 Cornell, E.: Byggnadstekniken – metoder och idéer genom tiderna. Byggförbundet 1970.

Fig 1.12 http://en.wikipedia.org/wiki/Dee_bridge_disaster

Fig. 1.21 Berridge, P.S.A.: The girder bridge after Brunel and others. Robert Maxwell Publisher, London 1969.

Fig. 2.2 Haunted Ohio. Horror for the holidays. Ghosts of the Ashtabula Bridge Disaster. www.prairieghosts.com/rr_disaster.html

Fig. 2.3 The Ashtabula Disaster. Harper's weekly – January 20, 1877. www.catskillarchive.com/rrextra/wkasht.html

Fig. 2.4 The Ashtabula Bridge Disaster. http://home.alltel.net/arhf/bridge.htm

Fig. 3.2 Wolcott, B.: The breaks of progress. Mechanical Engineering, March 2004, Vol. 126, Issue 3, pp. 32–35.

Fig. 3.7 Wolcott, B.: The breaks of progress. Mechanical Engineering, March 2004, Vol. 126, Issue 3, pp. 32–35.

Fig. 3.18 Tom Martin's Tay Bridge Disaster Web pages. http://taybridgedisaster.co.uk/

Fig. 3.21 Tay Bridge, Dundee. www.mccrow.org.uk/TaysideToday/TayBridges/TayBridges.htm

Fig 3.22 Yapp, N.: 150 Years of Photo Journalism. Volume I. Könemann, Köln, 1995.

Fig. 3.23 The Forth Bridge (post-card).

Fig. 4.8 Hammond, R.: Engineering Structural Failures – the causes and results of failure in modern structures of various types. Odhams Press Limited, Long Acre, London, 1956.

Fig. 4.13 The First Quebec Bridge Disaster – A Case Study. www.civeng.carleton.ca/ECL/reports/ECL270/Introduction.html

Fig. 4.24 Hammond, R.: Engineering Structural Failures – the causes and results of failure in modern structures of various types. Odhams Press Limited, Long Acre, London, 1956.

Fig. 4.25 Hammond, R.: Engineering Structural Failures – the causes and results of failure in modern structures of various types. Odhams Press Limited, Long Acre, London, 1956.

Fig. 4.26 The Quebec Bridge. http://www.civeng.carleton.ca/Exhibits/Quebec_Bridge/intro.html

Fig. 4.27 104 – Le Pont de Québec. Quebec Bridge (postcard). Lorenzo Audet ENR., Éditeur, 1226 Ave. des Pins, Qué., P.Q.

Fig. 5.1 Stålbygge. Utdrag ur Stålbyggnad, programskrift 11 från Statens råd för byggnadsforskning, Svenska Reproduktions AB, Stockholm 1971.

Fig. 6.2 Gudmundrå vykort – Sandöbron. http://www.famgus.se/Gudmund/Gudm-Sandobron.htm

Fig. 6.10 Reference missing.

Fig. 6.11 Sandöbron (Postcard, Pressbyrån).

Fig. 7.4 Tacoma Narrows Bridge Photograph Collection. Special Collections, University of Washington, Libraries, Seattle, Washington. "Tacoma Narrows Bridge during collapse showing central span twisting". UW21413. http://www.lib.washington.edu/specialcoll/

Fig. 7.5 Tacoma Narrows Bridge Photograph Collection. Special Collections, University of Washington, Libraries, Seattle, Washington. "Film still showing car tilting to the right in twisting Tacoma Narrows Bridge, November 7, 1940". UW21429. http://www.lib.washington.edu/specialcoll/

Fig. 7.6 Tacoma Narrows Bridge Photograph Collection. Special Collections, University of Washington, Libraries, Seattle, Washington. "Tacoma Narrows Bridge midsection collapsing into the waters of the Tacoma Narrows, November 7, 1940". UW21422. http://www.lib.washington.edu/specialcoll/

Fig. 7.7 Tacoma Narrows Bridge Photograph Collection. Special Collections, University of Washington, Libraries, Seattle, Washington. "Side Girder rising while roadbed falls". UW27459z. http://www.lib.washington.edu/specialcoll/

Fig. 7.8 Tacoma Narrows Bridge Photograph Collection. Special Collections, University of Washington, Libraries, Seattle, Washington. "Aerial view of collapsed 1940 Narrows Bridge". UW26818z. http://www.lib.washington.edu/specialcoll/

Fig 7.17 Tacoma Narrows Bridge Photograph Collection. Special Collections, University of Washington, Libraries, Seattle, Washington. "Completed Current Narrows Bridge as viewed from west side, Gig Harbor, 1950". UW7091. http://www.lib.washington.edu/specialcoll/

Fig. 8.4 Peace River Bridge in 1950. http://www.explorenorth.com/library/akhwy/peaceriverbridge.html

Fig. 8.5 Suspension Bridge Terminology. BridgeSpeak, Featuring the Peace River Bridge. Cable Band. http://www.inventionfactory.com/history/RHAbridg/term/

Fig. 8.6 Suspension Bridge Terminology. BridgeSpeak, Featuring the Peace River Bridge. Tower Saddle. http://www.inventionfactory.com/history/RHAbridg/term/

Fig. 8.7 Suspension Bridge Terminology. BridgeSpeak, Featuring the Peace River Bridge. Cable Bent. http://www.inventionfactory.com/history/RHAbridg/term/

Fig. 8.8 Suspension Bridge Terminology. BridgeSpeak, Featuring the Peace River Bridge. Anchor. http://www.inventionfactory.com/history/RHAbridg/term/

Fig. 8.9 Suspension Bridge Terminology. BridgeSpeak, Featuring the Peace River Bridge. Anchor. http://www.inventionfactory.com/history/RHAbridg/term/

Fig. 8.11 The Fort St. John – North Peace Museum. http://collections.ic.gc.ca/north_peace/transport/02.07.html

Fig. 9.3 McGuire, W.: Steel Structures. Prentice-Hall, Inc. / Englewood Cliffs, N.J., 1968.

Fig. 9.4   McGuire, W.: Steel Structures. Prentice-Hall, Inc. / Englewood Cliffs, N.J., 1968.

Fig. 10.7   Hopkins, H.J.: A span of bridges – an illustrated history. Praeger Publishers, New York 1970.

Fig. 11.2   Dicker, D.: Point Pleasant Bridge collapse mechanism analyzed. Civil Engineering (ASCE), July 1971.

Fig. 11.3   Silver Bridge. http://en.wikipedia.org/wiki/Silver_Bridge

Fig. 11.5 Anglesey – Mon Info Web. http://www.anglesey.info/Menai%20 Bridges.htm

Fig. 12.5   Chalmers University of Technology, Department of Structural Engineering, Steel and Timber Structures, Göteborg (slide picture of the Fourth Danube Bridge).

Fig. 12.10   Aurell, T. – Englöv, P.: Olyckor vid montering av stora lådbroar. Väg- och vattenbyggaren 5, 1973.

Fig. 12.11   Aurell, T. – Englöv, P.: Olyckor vid montering av stora lådbroar. Väg- och vattenbyggaren 5, 1973.

Fig. 13.3   Ryall, J.M.: Britannia Bridge, North Wales: Concept, Analysis, Design and Construction. Proceedings of the International Historic Bridge Conference, Columbus, Ohio, USA, August 27–29, 1992.

Fig. 13.4   Hopkins, H.J.: A Span of Bridges – an illustrated history. Praeger Publishers, Inc., New York, 1970.

Fig. 13.7   Werner, E.: Die Britannia- und Conway-Röhren-brücke. Werner-Verlag, Düsseldorf, 1969.

Fig. 13.8   Robert Stephenson and Britannia Tubular Bridge over the Menai Straits. http://vivovoco.rsl.ru/VV/E_LESSON/BRIDGES/BRIT/BRIT.HTM

Fig. 13.11   Conway Castle from Gryffin Hill (post-card).

Fig. 13.13   2 D 5 3 Britannia Bridge (Fire report pictures) http://www.2d53.co.uk/ britanniabridge/menu.htm

Fig. 13.14   Typical fracture in tubes after the fire. Unknown reference. Material provided by late Bernard Godfrey, Imperial College, London.

Fig. 13.15 Stephenson and Britannia Tubular Bridge over the Menai Straits. http://vivovoco.rsl.ru/VV/E_LESSON/BRIDGES/BRIT/BRIT.HTM

Fig. 13.17 Stephenson and Britannia Tubular Bridge over the Menai Straits. http://vivovoco.rsl.ru/VV/E_LESSON/BRIDGES/BRIT/BRIT.HTM

Fig. 14.5   Aurell, T. – Englöv, P.: Olyckor vid montering av stora lådbroar. Väg- och vattenbyggaren 5, 1973.

Fig. 15.5   Report of Royal Commission into the Failure of West Gate Bridge. The West Gate Bridge Royal Commission Act 1970, No. 7989, Victoria, 1971. VPRS No. 2591/PO, Unit 14, Transport of Proceedings.

Fig. 15.6   Report of Royal Commission into the Failure of West Gate Bridge. The West Gate Bridge Royal Commission Act 1970, No. 7989, Victoria, 1971. VPRS No. 2591/PO, Unit 14, Transport of Proceedings.

Fig. 15.15   Report of Royal Commission into the Failure of West Gate Bridge. The West Gate Bridge Royal Commission Act 1970, No. 7989, Victoria, 1971. VPRS No. 2591/PO, Unit 14, Transport of Proceedings.

Fig. 15.17   Paul, A.M.: När den stora bron rasade. Det Bästa, October 1972. Summary (in Swedish) from Sunday Independent, 18 June 1972.

Fig. 15.18   Report of Royal Commission into the Failure of West Gate Bridge. The West Gate Bridge Royal Commission Act 1970, No. 7989, Victoria, 1971. VPRS No. 2591/PO, Unit 14, Transport of Proceedings.

Fig. 15.19   Report of Royal Commission into the Failure of West Gate Bridge. The West Gate Bridge Royal Commission Act 1970, No. 7989, Victoria, 1971. VPRS No. 2591/PO, Unit 14, Transport of Proceedings.

Fig. 15.20   Some removed and saved parts from the collapsed West Gate Bridge, at the Department of Civil Engineering, Monash University, Melbourne (with kind permission of prof. Em. Bo Edlund, Chalmers University of Technology, Göteborg, Sweden).

Fig. 16.5   Chalmers University of Technology, Department of Structural Engineering, Steel and Timber Structures, Göteborg (slide picture of the Rhine Bridge in Koblenz).

Fig. 17.3   Ekardt, H-P.: Die Stauseebrücke Zeulenroda. Ein Schadensfall und seine Lehren für die Idee der Ingenieurverantwortung. Stahlbau 67 (1998), Heft 9. With kind permission of Karl-Eugen Kurrer, Editor-in-chief, Stahlbau.

Fig. 17.4   Ekardt, H-P.: Die Stauseebrücke Zeulenroda. Ein Schadensfall und seine Lehren für die Idee der Ingenieurverantwortung. Stahlbau 67 (1998), Heft 9. With kind permission of Karl-Eugen Kurrer, Editor-in-chief, Stahlbau.

Fig. 17.5   Ekardt, H-P.: Die Stauseebrücke Zeulenroda. Ein Schadensfall und seine Lehren für die Idee der Ingenieurverantwortung. Stahlbau 67 (1998), Heft 9. With kind permission of Karl-Eugen Kurrer, Editor-in-chief, Stahlbau.

Fig. 17.6   Ekardt, H-P.: Die Stauseebrücke Zeulenroda. Ein Schadensfall und seine Lehren für die Idee der Ingenieurverantwortung. Stahlbau 67 (1998), Heft 9. With kind permission of Karl-Eugen Kurrer, Editor-in-chief, Stahlbau.

Fig. 18.1   Permanent monitoring of Viennese Reichsbrücke. History of Reichsbrücke. http://www.reichsbruecke.net/geschichte_e.php

Fig. 18.2   Permanent monitoring of Viennese Reichsbrücke. History of Reichsbrücke. http://www.reichsbruecke.net/geschichte_e.php

Fig. 18.3   Reiffenstuhl, H.: Collapse of the Viennese Rechsbrücke: Causes and Lessons. IABSE Symposium, Washington, 1982, Final Report.

Fig. 18.5   Permanent monitoring of Viennese Reichsbrücke. History of Reichsbrücke. http://www.reichsbruecke.net/geschichte_e.php

Fig. 18.6   Permanent monitoring of Viennese Reichsbrücke. History of Reichsbrücke. http://www.reichsbruecke.net/geschichte_e.php

Fig. 18.12   Reiffenstuhl, H.: Collapse of the Viennese Rechsbrücke: Causes and Lessons. IABSE Symposium, Washington, 1982, Final Report.

Fig. 18.14   Der Reichsbrückeneinsturz. http://www.wien-vienna.at/geschichte.php?ID=737

Fig. 18.15   Permanent monitoring of Viennese Reichsbrücke. History of Reichsbrücke. http://www.reichsbruecke.net/geschichte_e.php

Fig. 19.3   Tjörnbron. Thuviks förlag, Skärhamn (postcard).

Fig. 19.7   Jättetankern "THORSHAMMER" på 228,250 ton passerar Almöbron. Carlaförlaget Lysekil (postcard).

Fig. 19.8   Tjörnbron. Vägverket (Sune Brodin), Adlink, 1984.

Fig. 19.17   Tjörnbron. Foto: Jan-Olov Montelius (postcard).

Fig. 20.3   The Latchford Bridge Failure (2003) http://www.thekingshighway.ca/latchford.html

Fig. 20.15   Latchford Bridge Collapses. Speaker, November 1, 2006. http://speaker.northernontario.ca/content/Feature%20Story/C%20Front%2011:01:06.pdf

Fig. 20.17   Photo of the Aubrey Cosens Bridge at Latchford taken in September 2007 by Cameron Bevers.

Back cover   Paul, A.M.: När den stora bron rasade. Det Bästa, October 1972. Summary (in Swedish) from Sunday Independent, 18 June 1972.

Index

aerodynamic instability (Chapter 7)
ageing 82, **85**, 87, 135
Alaska 115
Alaska Highway 116, 122
Albert Bridge 99, 114
Albert Canal 79
Almö Bridge **223**
Almön 223
Ammann, O. 103, 111
anchorage failure (Chapter 8)
Angers Bridge 98, 114
angle block **17**
Anglesey 2, 155–156, 166
Archduke Franz Ferdinand 213
Ashtabula Bridge **17**, 37, 40, 73, 99
Askerö Sound 223, 225–226
Aubrey Cosens Bridge **235**

Baker, B. 37
Balmoral Castle 36
basic Bessemer 85
Berlin 80
Berlin Wall 201
Bessemer 85–86
Birmingham 1
Bohuslän 223
Bouch, T. 33–34, 36–39, 47–48, 50–51
Bouscaren 67, 74
Brighton Pier 98
Britannia Bridge 2, 53, **155**, 171, 190, 194
British Columbia 115, 123
brittle fracture (Chapter *1, 2, 3, 5, 10, 11, 20*)
Bronx-Whitestone Bridge 111
Brooklyn (Suspension) Bridge 50, 99, 140
Brücke der Roten 215
Brussels 79

buckling (Chapter *4, 6, 9, 12, 14, 15, 16, 17, 19*)
Buffalo 17
Burchell, Mr. 18
Burrard Inlet 123

Canadian Government 58
cast iron (Chapter *1, 2, 3*)
chain links 3
Chalmers University of Technology 92
Charpy impact test 82–84, 136
Charpy, G. 82
Chester 2–3, 8, 155, 164
Chicago 17–18
Chicago Tribune 19
Cleddau Bridge **171**, 182, 189, 194, 211
Cleveland 17, 26
Coalbrookdale 1
Collins, C. 24
composite action (Chapter *4*)
Conway 164, 169
Conway Bridge 159, 164, 169
Cooper, T. **53**
corrosion 9, 87, 145–146
corrosion fatigue 146
cotters 39–41
Cox, J. 34
Cremona diagram 57
critical (crack) length 11, 28, 30, 80, 87, 146, 240
Culmann diagram 57
Current Tacoma Narrows Bridge 112–113
Czech Republic 201
Czechoslovakia 201

damping 42, 110–111
Danube River 149, 221
Dawson Creek 115–116

Dee Bridge **1**, 18, 21, 29
Dicker, D. 146
Diver, The 50
Dolomite rock 85
ductility 28, 30, 54, 85, 136
Dundee **33**

Eads Bridge 53, 55
Eads, J. 53
East River 75, 111, 140
Edinburgh 33–34
Edmonton 115
Eiffel Tower 55
Eiffel, G. 55
Ellet, C. 99–100
Empire Bridge 213
Euler buckling 72, 125, 197
Euler theory 65

Fairbairn, W. 162–163
fatigue (Chapter *2, 3, 10, 11, 20*)
Ferdinand, Archduke Franz 213
fire (Chapter *13*)
First World War 213
Firth of Forth 33–34, 48, 50
Firth of Tay **33**
Fitzmaurice, M. 75
Forssell, C. 92
Fort St. John 115–116
Forth Bridge 37, 49, 55–56, 58, 60–61,
    73–75, 78, 81, 111, 232
Forth Suspension Bridge (Bouch's design) 48
Fourth Danube Bridge **149**, 171–172, 210,
    215
Fowler, J. 37
Fraser River 123
Freeman, Fox & Partners 172, 179, 191
fretting 143–144
Freyssinet 91
friction 28, 117, 143–144, 146, 163, 245

Galloping Gertie 100
George Washington (Suspension) Bridge
    99–100, 103
Gerber beam 57–58
Gerber, H. 57
Golden Gate Bridge 99–100, 103
Göta Älv 54
Göteborg 92, 223–224
Gothenburg 92, 223–224
Granholm, H. 92, 94

Great Lakes Region 17, 53, 235

Hardenberg Strasse 80, 85
Harper's Weekly 20, 29
Hasselt (Road) Bridge **79**
Hercilio Luz Bridge 148
Herentals Oolen 80
hit damage (Chapter *19*)
Hodgkinson, E. 162–163
Holyhead 2, 155, 164
Howe Truss **17**
Howe, W. 21
Hutchinson, General 39

inspection (Chapter *2, 11, 20*)
impact transition temperature (ITT) 83–85
Ironbridge 1, 2, 15, 38
Ironworkers Memorial Second Narrows
    Crossing 128
Iroquois Indians 19

Jourawski, D.J. 161

Källön 223, 233
Kármán, T. von 103
Kaulille 80
Kings Bridge **129**
Koblenz 193–194, 201
Kronprinz-Rudolph-Brücke 213, 215
Kusfors Bridge 243–246

Lake Erie 17, 19
Lake Michigan 141
Lake Ontario 53–54
Lake Shore and Michigan Southern Railway
    Company 17
Latchford 235, 245
Latchford Bridge **235**
Leipzig 201
Lindenthal, G. 55, 75
Lower Yarra River 179

Mackinac Bridge 141
Main River 57
Maine River 98
Manhattan Bridge 75, 100
McLure 63, 73
Meccano 25
Melbourne 129, 179, 191, 201
Menai (Strait Suspension) Bridge 98, 139,
    141, 147, 168

Menai Strait 2, 53, 141, 155–156, 164
Milford Haven **171**, 180, 182, 189, 201
Millennium Bridge 113
Mississippi River 53
Modjeski, R. 75
Moisseiff, L. 100
Monash University 191
Montreal **53**, 164
Montreal River 235
Mosel 193

New Tay Bridge 49, 55
New York 17, 60, 63, 74–75, 99, 101, 111, 140
Newport 37
Niagara Falls Suspension Bridge 48
non-composite action (Chapter *4*)
North British Railways 33

Ohio 17–18, 139–140
Ohio River 99, 139
Ontario 235
ovalization 13–14

Peace River Bridge **115**, 123
Pearl Harbor 115
Philadelphia 55
Phoenix Bridge Company 55–56, 63–64, 74–75
Phoenixville 55
pin bolts 3, 13
Pittsburgh 99
plastic deformations 40, 86, 166
plate buckling (Chapter *12, 14, 15, 16, 17*)
plate thickness 82–84, **84**, 159
Plougastel Bridge 91
plugs 2
Point Pleasant Bridge **139**, 214, 219, 245
Prater Bridge 154
Pratt truss 163
prestressed concrete 91, 221
pretensioning 4, 7, 28
Prince of Wales (the steam vessel) 162
Puget Sound 97, 101

Quebec Bridge 53, 100, 111, 124
Quebec Bridge Company *53, 55*, 75
Queen Post truss 2–4, 7, 14–15
Queen Victoria 36, 53
Queensboro Bridge 75

Red Army Bridge 215
Red Rock Bridge 55
redundancy 22, 147, 243
Reichsbrücke **213**, 245–246
resonant vibrations 42, 44, 105
Rhine Bridge **193**
Rio de Janeiro 113
Rio-Niterói Bridge 113
rivets 2, 61, 63, 72, 74, 163
robustness 21–22, 24, 29, 66, 74–75, 99, 136, 154, 164, 234, 243, 245
rocker bearing 245–246
rocker tower 141, 214
Rocket, The 155
Roebling Company 99
Roebling, J.A. 48, 99
Royal Suspension Chain Pier 98
Ruabon 8–9
Rüdersdorf 80, 84–85, 137

St. Lawrence River 53–54, 56, 61, 75, 78, 164
St. Louis 53
St. Marys Bridge 148
Sandö Bridge **89**, 223, 243
Sarajevo 213
scour 39, 118
Second Narrows Bridge **123**
Second World War 92, 164
Sgt. Aubrey Cosens VC Memorial Bridge **235**
Severn Bridge 113, 172
Severn River 1, 38, 113
Shepherd, M. 19
Shropshire 1
Silver Bridge **139**
Silver Memorial Bridge 148
Skellefte Älv 243
sliding friction 143–144, 146
splitting failure (Chapter *18*)
Stahlbau 201–204
Stålpilen 243–244
Star Clipper 224, 228, 234
steel 6, 11–12, 14–15, 25, 30, 49, 53–54, 56, 58, 60, 70, 80–81, 85–88, 99, 101, 129, 135–136, 139–140, 143, 146, 155, 161, 232, 240
Steinman Type 141, 214
Steinman, D.B. 101, 141, 148
Stenungsund 223
Stephenson, G. 53, 155

Stephenson, R. 2, 9, 53, 155, 157, 161–164, 166, 168–169, 190
Stockholm 89
Stone, Amasa 19, 21, 24–25, 31
Stone, Andros 24
Strait of Georgia 123
stress corrosion 146
Südbrücke 193
Swan, Wooster & Partners 126
Swedish Maritime Administration 226
Sydney Harbour Bridge 179

Tacoma Narrows Bridge **97**, 121, 141
Tay Bridge **33**, 53, 56, 73, 75, 219
Tay Bridge, New 48–49, 55
Taylor 115–116
Telford, T. 98, 168
Thames River 99, 114
Thomas steel 85–87
Thorshammer 223, 227
Tjörn 223, 234
Tjörn Bridge 223, 233–234
Trent Valley 15

Uddevalla 223, 226

Vancouver 123
Vänersborg 54
Vargön 54
Victoria Bridge 53, 164

Victoria Cross 235
Vienna 149, 201, 213, 215
Vierendeel, A. 79–80, 87
von Kármán, T. 103
vortex shedding 105–106, 109–111, 113

Wales 2, 53, 98, 113, 139, 141, 155–156, 171
Watson, S.R. 101
web buckling (Chapter 9)
welding 81, 82, **86**, 87–88, 131, 133, 135, 137, 152, 195, 199–200, 210
West Gate Bridge **179**, 193–195, 200, 211
West Virginia 99, 139–140, 148
Wheeling Suspension Bridge 99–100
wind gusts 37, 42, 44
wind loading (Chapter 3, 7)
Windsor 37
Woodruff, G.B. 103
wrought iron 6, 9, 14–15, 50, 54, 161–163, 166, 168
Wye Bridge 172

Yarra River 129
Yolland, W. 38–39

Zeulenroda Bridge 152, **201**

Ångermanälven 89, 95